Hotels and Resorts

Hotels and Resorts: an investor's guide presents a comprehensive analysis of how hotels, golf courses, spas, serviced apartments, gyms, health clubs and resorts are developed, how they operate and how they are valued. Drawing on over eighteen years' experience in the leisure property industry, David Harper provides invaluable advice on how to buy, develop and sell such properties. Working through the required due diligence process for purchases (including how to identify a 'good buy'), through the 'route map' for a successful development and ending with how to ensure you maximise your returns when selling the asset, this book covers the whole lifecycle of leisure property ownership.

Examples of valuations, development issues and sales processes taken from the USA, UK, France, Nigeria, Kenya, Australia, Hong Kong, Singapore and Brazil provide in-depth analysis on the similarities and differences in approach to hotels and resorts in various parts of the world.

This book provides invaluable guidance to international investors, developers, asset managers and students in related subject areas.

David Harper is the managing director of Leisure Property Services, a fellow of the Royal Institute of Chartered Surveyors and the author of *Valuation of Hotels for Investors*. He is a founding member of Hotel Partners Africa (HPA), a consultancy offering a comprehensive range of services to hotel owners, operators and investors from initial feasibility to sale and exit.

Hotels and Resorts

An investor's guide

David Harper

LONDON AND NEW YORK

First published 2017
by Routledge
2 Park Square, Milton Park, Abingdon, Oxon OX14 4RN

and by Routledge
711 Third Avenue, New York, NY 10017

Routledge is an imprint of the Taylor & Francis Group, an informa business

British Library Cataloguing in Publication Data
A catalogue record for this book is available from the British Library

Library of Congress Cataloging-in-Publication Data
Names: Harper, David (Property services manager), author.
Title: Hotels and resorts : an investor's guide / David Harper.
Description: Abingdon, Oxon ; New York, NY : Routledge, 2017. |
Includes bibliographical references and index.
Identifiers: LCCN 2016005445 | ISBN 9781138690158 (hardback : alk. paper) | ISBN 9781138853744 (pbk. : alk. paper) | ISBN 9781315722610 (ebook : alk. paper)
Subjects: LCSH: Hotels--Valuation--Handbooks, manuals, etc. |
Resorts--Valuation--Handbooks, manuals, etc.
Classification: LCC TX911.3.V34 H37 2017 | DDC 910.46--dc23
LC record available at http://lccn.loc.gov/2016005445

ISBN: 978-1-138-69015-8 (hbk)
ISBN: 978-1-138-85374-4 (pbk)
ISBN: 978-1-315-72261-0 (ebk)

Typeset in Goudy
by HWA Text and Data Management, London

This book is dedicated to Arabella, Nicola and Matthew

Contents

Figures

Tables

Preface

Welcome to *Hotels and Resorts: an investor's guide*. The purpose of this book is to act as a guide to all those keen to get involved in the hospitality market, whether as developers, owners, investors, operators or advisors.

The book is divided into three main sections. The first part outlines the investment lifecycle for a hotel, resort or leisure property. It provides an introduction to hotels and resorts, providing commentary on who invests in such properties, as well as why they invest. It provides an 'ownership guide', outlining the various types of ownership of such properties, including the various legal agreements common in the industry, as well as outlining 'asset management' in terms of the hospitality industry. It provides a 'buyers' guide', which discusses what makes a good buy, and risk versus reward in terms of pricing and yield selection. The 'development guide' outlines the 'development road map', discusses the need for a feasibility study, outlines the key criteria involved in operator selection and discusses the main components involved in designing a hotel or resort. The 'construction guide' outlines how to identify, understand and manage the risk inherent in the construction process. Finally, Part I concludes with the 'selling guide', which explains the disposal process, discusses enhancing value and advises on how to maximise the price achieved during the sale.

Part II discusses the valuation process for hotels and resorts specifically, though the key points are equally valid for all types of leisure properties. It outlines the key methods of valuation, providing simple examples. The importance of due diligence is discussed, outlining the various types of due diligence including legal, statutory, condition and environmental. The chapter on financial due diligence explains the importance of understanding the detail of the business. It provides a number of worked examples on how to analyse the operation, so accurate projections for future performance can be made. Valuation due diligence is discussed, highlighting the inspection process, assessing the local hotel market, the interview with the general manager and finally how to analyse all the data.

The final section provides examples of specific types of valuations. The details from Part II on due diligence apply to all of these individual chapters, but each chapter discusses specific issues arising from that type of property. Property types outlined include spas, gyms and golf courses, all of which can be stand-alone leisure properties or can form part of a hotel or resort. In addition, other chapters cover serviced apartments, fractional ownership, site valuation and rental valuations.

Throughout the book, text boxes are included from some of the most respected people in the market, giving advice, warnings or just recounting tales of actual events, characters or issues that have been encountered within the industry.

This book can be read in the same way as a novel, and indeed for those new to the industry wishing to get an easy introduction into the market, this is the recommended approach to the book. However, many readers will have a specific interest, or be looking for something in particular, and in this case referring just to one chapter or part of the book is just as sensible.

The hotel world is full of jargon and 'technical speak', and it has not been possible to eliminate all jargon from the text of this book. The first time a technical word is used it has been explained which is useful for those reading the book cover to cover. Each term is also included in the glossary, so those dipping in and out can also find explanations of unfamiliar words and terms.

In the interests of confidentiality, while all the examples are based on real cases, the names of the hotels and resorts have been fictionalised and resemblance to the name of any actual hotel or resort is purely coincidental.

I do hope you enjoy the book.

Part I
The investment lifecycle

This section of the book details all the ownership phases for a hotel investor, outlining ownership issues, buying criteria, development and construction parameters, as well as the oft-forgotten phase: the exit.

The intention is to allow all those involved in the hospitality industry to review the typical phases of the investment lifecycle, as well as allowing potential developers, purchasers, owners or sellers to review the individual chapters of direct interest to them.

The first chapter provides an introduction to hotels and resorts, providing commentary on who invests in such properties, as well as why they invest.

The second chapter provides an 'ownership guide', outlining the various types of ownership of such properties, including the various legal agreements common in the industry. In addition, it reviews 'asset management' in terms of the hospitality industry.

The third chapter provides a 'buyers' guide', which discusses what makes a good buy, including risk versus reward, in terms of pricing and yield selection.

The fourth chapter is the 'development guide' and outlines the 'development road map', discusses the need for a feasibility study, outlines the main criteria involved in operator selection and discusses the key components involved in designing a hotel or resort.

The fifth chapter comprises the 'construction guide' and outlines how to identify, understand and manage the risk inherent in the construction process.

Finally, Part I concludes with a chapter dealing with disposal. The 'selling guide' explains the disposal process, discusses enhancing value and advises on how to maximise the price achieved during the sale.

1 Introduction to hotels, resorts and leisure property buyers

What is a hotel/resort?

This book is primarily about the hospitality industry and covers hotels, resorts and associated leisure properties. As such, it is important to define what we mean by these terms. Probably the easiest place to start with is 'hotel', as the vast majority of people think they already know what a hotel is. According to the Oxford English Dictionary, a hotel is 'an establishment providing accommodation, meals, and other services for travellers and tourists, by the night'. Wikipedia, on the other hand, defines a hotel as 'an establishment that provides lodging paid on a short-term basis'. As you can see from these two sources there is not a consistent definition of what a hotel actually is, with the Oxford English Dictionary definition talking more about full-service hotels, while Wikipedia uses a 'short-term' definition, which excludes the whole of the extended-stay part of the market.

What is agreed by all is that a hotel provides a bed for a guest to stay for the night (or longer). That bed can be in a shared room or a private room, with or without an exclusive bathroom, and with or without breakfast or other food options. The hotel could have meeting spaces, leisure facilities, car parking and other facilities. It could be of a luxury standard, or provide only the most basic letting accommodation. The letting period can be for a number of hours or for an extended period. These are all differences between various market segments and not things that define a hotel.

A hotel is not a private apartment let for over six months at a time (effectively a long-term contract), although hotel rooms can be let for periods longer than this time. A 'couch-share' is not a hotel room either, albeit this relatively new market is having an impact on the profitability in some segments of the hotel market.

A resort differs slightly from a hotel, insofar as it usually has more facilities than a stand-alone hotel. These facilities provide part of the attraction to guests, and are an integral part of why the booking is made. Most resorts will have various standards of accommodation available to guests, along with a variety of facilities, for example, swimming pools, restaurants, bars, tennis courts, a golf course, a spa, a beach bar or a conference centre.

The importance of hotels

Since the 1960s, hotels have been becoming more and more important as economic drivers across the world. The US hospitality industry has typically led the way with Europe, then Asia and South America, following, and, in recent years, with Africa following on strongly. The increase in supply has been dramatic but for the most part it has been market-led, as a change in the economic environment has generated a greater demand for hotel rooms across large parts of the world.

There are many reasons why the hotel industry is important to the wider economy.

Changing dynamic of demand

As business has required people to move further and further afield, the provision of hotel accommodation has been essential to the smooth running of the worldwide economy. International conglomerates requiring places where staff can meet up, accommodation for travelling sales people, convenient accommodation for inconveniently timed flights, conference hotels for networking events, places to stay while visiting friends and family or luxury resorts to unwind for a well-earned holiday have all enhanced demand for hotels.

Arguably, as an economy changes its structure from a primary economy (farming and mining) to a secondary economy (manufacturing), through to a tertiary (service-led) economy, the level of demand for hotel accommodation has increased throughout the cycle. Indeed, eventually, as the economies mature, there are examples where people are staying in a hotel purely for the experience of staying there, rather than for any practical reason.

As an investment class

The size of the worldwide hotel industry is estimated to have increased significantly over the last 100 years, and continues to grow at different rates in different regions. According to STR Global, in the ten years up to 2014, the number of hotel bedrooms grew by 1.03 per cent in the Americas, 3.61 per cent in Asia-Pacific, 1.02 per cent in Europe and 2.68 per cent in the Middle East and Africa.

As an employer

Hotels employ a wide range of people across all segments of society, from professional staff to unskilled labour, thereby providing opportunities for the local population to enhance their economic standing through hard work and training. As the size of the hotel market increases, so does its impact on direct job creation. In addition, it generates secondary employment in complementary industries (for example, laundry services, taxi services, etc.). It is estimated that for every direct job created in the hotel industry, an additional eight jobs are sustained through indirect means (taxi drivers, waiters and chefs, bar staff, laundry workers, staff at suppliers, etc.), ensuring that hotels are one of the most effective ways to enhance employment prospects in the local area.

Earning foreign currency

Many hotels cater to non-domestic guests and have the ability to earn much needed foreign currency, helping governments balance their foreign exchange revenues.

Generating taxable revenues

A hotel generates significant amounts of taxable revenue for the tax authority if it is well located. Income tax from the staff, company tax on the profit of the business, property tax, licence fees and capital gains tax (when the property is sold) are the key taxes, making hotels a key contributor to local and national revenues.

As a place to do business

One of the first steps for most newly developing countries is to develop a quality hotel, as without one there is limited opportunity to attract international developers and investors into the country to discuss business opportunities.

Enhancing the status of a country in the eyes of the world

A quality hotel can help enhance the perception of a country to other nationalities who know little about that country.

What actually is real estate?

Hotels, resorts and leisure properties are all real-estate investments. Real estate (sometimes called 'real property' or even just 'property') consists of land and/or buildings. The capital value attached to the real estate as an investment is directly linked to the transferability of the ownership interest in the property. If an interest in land and/or a building cannot be transferred to another party, then there is essentially no capital value attached to that interest, although the use of the site might be worth something to the current occupier.

All around the world there are many ways to hold real estate and, in simple terms, the less restrictions on the ownership of the land, the more that particular interest in the land is worth. Land law varies but in essence, the more you can do with the land, the longer it is owned for and the easier it is transferred, the more it will be worth.

As such, freehold title is usually preferred by most investors to leasehold title, which in turn is considered to be more desirable than commonhold land and licences.

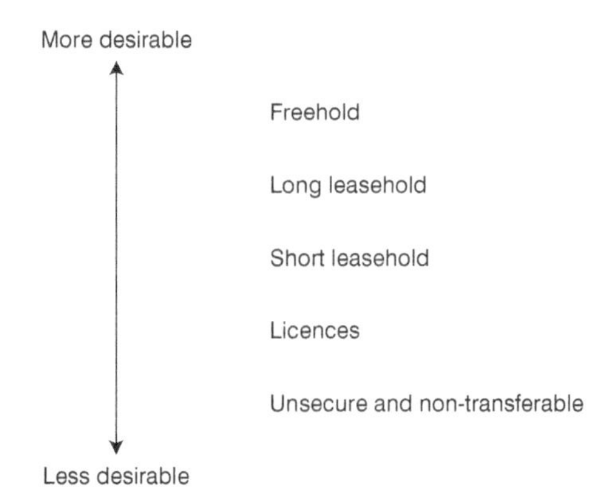

Figure 1.1 Desirability of tenure

Land law varies across the world but essentially, leasehold title transfers the right to exclusively use the property for a specific period and over that period of ownership, restrictions can be placed upon the use and enjoyment of the land and/or buildings.

Commonhold is relatively prevalent for residential properties in Australia and the US, but in terms of resorts, it tends only to be seen when looking at mixed-use resorts where timeshare or fractional ownership has been introduced. Commonhold effectively replicates freehold title for multi-occupancy buildings where traditionally only leasehold title would have been possible.

Licences, on the other hand, allow for the use of a property/building on a non-exclusive basis, which generally means that it has limited transferability (and therefore value).

In terms of a hotel investment, the 'property' being transferred is effectively the land, the building(s) and the operational business (or the benefits of an operational lease).

Appeal of real estate as an investment

In simple terms, an investment is somewhere that money is placed to either protect the capital invested or to earn a return, or ideally both. Real estate is one class of investment that can provide both an income (rent received) as well as capital growth. Other investment options are extensive and include items as diverse as government bonds, savings accounts, art, fine wines, stocks and shares.

An investor will choose where to invest their money based upon the specific criteria of that investment and how the options suit their particular requirements. In general, it is true to say that the less risky the investment, the lower the investment returns, and the balance between risk and reward is usually a key investment consideration.

Property has been a very popular investment for a long time because it is considered an area where, in the long term, capital values should rise due to the limitations on supply and perceived long-term increases in demand. Certainly in the UK, residential properties have seen substantial rises in capital value over the last 100 years, led by the increased demand for properties. This is partly down to population increase, partly down to the breakdown of traditional family units and partly down to the increasing availability of borrowing facilities. At the same time, there has been a lower increase in the supply of such properties due to a number of factors, not least various planning restrictions and lack of available land.

Pension funds and other investment institutions have invested in real estate to diversify their investment portfolios. The overall percentage of money invested in property by such a fund varies year on year, based upon the current thinking of the requirements for such investment vehicles. Traditionally though, such investment tended to be limited to retail, commercial and industrial property investments as they were seen as safer than other asset classes and there was a strong and measurable market.

Investment characteristics of real property

Real property has some key attributes that define its relative attractiveness to various different types of investors. Below they are divided into what are generally seen as positive attributes and negative attributes.

Positive attributes

Security

The relatively consistent demand, along with the perception that 'wealthy' people own real estate, ensures it has a consistent popularity as an investment class among individual investors.

Tangibility

The 'physical' nature of real estate with a 'tangible presence' provides comfort to a certain group of investors.

Enhancing value through asset management

Property is one of the few investment classes where the owner can actively enhance value through their own activities, whether through good asset management or planning gain, or by changing the occupier or occupier mix.

Figure 1.2 The Palm in Dubai

Scarcity

As Mark Twain once said, 'Buy land, they're not making it anymore'. In an expanding economy, land remains a finite substance (except in places like the Palm Islands in Dubai or Eko Atlantic in Lagos, Nigeria where land has actually been created). This provides a certain intrinsic appeal to investors, with the logic being that as demand increases and supply remains static, value must rise.

Negative attributes

Illiquidity

Compared with other investment classes, real estate is relatively slow to transact, and the time delay in being able to release the capital investment can be a significant disadvantage for many types of investors.

High transaction costs

Real-estate transaction costs can be quite high compared with many other investment classes. Legal fees and valuation fees, along with property transaction taxes, mean that every transaction is a relatively expensive exercise for the investor. When an investor likes to regularly update their investment portfolio, such transaction costs can reduce investment returns significantly.

Figure 1.3 Property in Ghana (property provides tangibility)

Imperfect market

In most markets around the world, property transactions are usually slightly secretive affairs and there is rarely a single source of good quality, transparent information that investors can rely upon. This lack of data deters many potential investors from entering the market, as other potential investments have more transparent market data.

Expensive and time consuming to manage

Although there is the potential to enhance value through asset management, the other side of this is that the cost of asset management (even where no enhanced value is possible) can be relatively expensive for real estate when compared to other investment classes.

Large lot sizes

Typically, real estate is a substantial investment for most investors and along with the inflexibility of lot sizes, this prevents many classes of investors from getting involved in the market.

These shortcomings can be substantial and so to reduce their impact, new ways to invest in property that minimise some of these negative attributes have been created.

These ways include indirect property investment in REITs (real estate investment trusts), shares in property companies, insurance property funds, property unit trusts and offshore property companies. These all have their own investment characteristics which help determine which type of investors they appeal to.

Characteristics of a hotel investment compared with property in general

In recent years, there has been a strong movement to diversify property holdings in many investors' property portfolios. Income-generating properties, such as hotels, have started to become popular with investors.

At the same time, the increase in the range of operators who are prepared to offer leases and management contracts has led to more options for potential hotel investors. This increasing demand has led to a wide variety of hotel investment structures being created, from EBITDA (earnings before interest, tax, depreciation and amortisation)-based leases, turnover leases, fixed RPI (retail price index) investment leases and management contract investments.

Negative attributes

A hotel tends to have all the same disadvantages as standard real-estate investments. They generally comprise large lot sizes. The largest single hotel purchase I was involved in cost over £500,000,000 for a single property!

Transaction costs remain high compared with other investment classes. The market for hotels is typically even more secretive than for standard real-estate investments, with ownership, prices and yields paid, as well as trading data, jealously guarded from public scrutiny.

In addition, a hotel has one additional deterrent to standard investors: the element of business risk associated with the asset class. Hotels are almost entirely dependent for their value on their earning capacity, so if there is a fundamental shift in the business profile of a hotel (or the business profile of its location), this can add another 'risk' to the potential investment.

This operational risk can be quite foreseeable, for example, a road closing taking potential customers away, or a new airport taking existing traffic away from another airport. Alternatively, some risk can be less foreseeable, for example, an outbreak of Ebola in West Africa perhaps, where there has never been any history of the disease. Risk may also come through a change in customers' demand patterns, for example, an expansion in couch-surfing at the expense of budget hotels.

Positive attributes

Hotels do have a number of advantages over other investment classes and, indeed, against more traditional property investment categories. These attributes attract specific investors who look for such characteristics in their investments.

Pros and cons of hotel investments

Pros

- Pride of ownership
- Potentially excellent returns
- Limited need for comprehensive refurbishment
- Returns can be enhanced by investor's management
- Great hedge against retail investments
- Complements mixed-use developments
- Relative 'security' from property investments
- 'Tangible' investment.

Cons

- Relatively large lot sizes
- Imperfect and non-transparent market
- Expensive and time consuming to market
- Relatively illiquid
- High transaction costs.

Pride of ownership

Hotels have a relatively unique appeal among property classes in providing true pride of ownership for the investor. Certain office buildings (Sears Tower, Empire State Building or Petronas Towers) and shops (Harrods, KaDeWe or Bloomingdales) have a similar appeal but the ability of hotels to generate such trophy status is unequalled by other property classes.

Potentially because of the ability of the owner to offer hospitality to potential clients and business partners ('please, come and stay at my hotel'), the amount of high-net-worth individuals (HNWIs) who own hotels is disproportionately large.

Potentially excellent returns

If the right property is built in the right place, is well managed and reacts to market demands, then hotels can generate excellent returns, above typical real-estate returns, because of the operational risk involved.

Furniture, fittings and equipment (FF&E) reserves

Hotels typically reserve a proportion of the revenue they generate into a sinking fund to keep the property in a good state of repair. This is because if the property deteriorates, then trading is negatively impacted as customers are unwilling to pay as high a price to stay in a shoddily decorated hotel in poor condition. As such, this standard investment means that hotels are one of the only asset classes where regular redevelopment is not required. The Lygon Arms in the Cotswolds, UK, was first opened in the 1530s as a hotel and has effectively remained in this property use since that time. Very few other commercial property investments can show even half a century of the same use.

Possibility to positively impact on returns through effective asset management

The very nature of the hotel business means that the operator needs to react swiftly to changes in the market dynamics if they are to maintain market share. The ability of the owner to significantly enhance their returns through their own actions is a key reason for investing in hotels, for certain investors.

A good hedge against retail investments

Retail remains the primary property investment class, with investment funds allocating significant proportions of their property investment into this class. Hotels have been shown to be an effective 'hedge' against the downturns in the retail investment market, making hotels a sensible way to mitigate risk for portfolio managers.

Complement mixed-use schemes well and help gain planning for wider valuable schemes

New property developments tend to be mixed-use schemes because the risk for the developer is inherently diversified. The authorities responsible for development control tend to favour such schemes because it brings variety to the area. Most importantly from the developers' perspective, values tend to be greater for such schemes if they have been well designed. Hotels form a very valuable part of most mixed-use schemes, as evening and night-time use bring a different dynamic to a larger development. Car parking facilities, for example, can be shared with other uses. Offices tend to require parking spaces only during normal work hours while hotels tend to need the majority of their spaces after working hours have finished. Demand for the hotel can be complemented by other uses, for example, business parks can generate direct demand for hotel use.

Who are the buyers?

One of the key questions most newcomers to the market ask is who is behind this influx of money into the hotel investment sector over the last thirty years and why has it happened? Is it just that the investment world has finally woken up to the attraction of the hotel market, or have the requirements of the investors changed, making it a much more attractive option than it used to be?

It is possible to categorise hotel investors as follows:

- public companies
- private companies
- pension funds

- investment funds
- insurance funds
- family funds
- private individuals
- governments and local authorities
- financial institutions
- partnerships
- high-net-worth individuals (HNWIs).

In fact, this is a simple list of most of the different types of entities and individuals that exist in any investment market, not just the hotel market. However, there is a higher interest from HNWIs and family funds than would usually be expected on a purely statistical basis.

Mixed-use properties – costs and service charges

Alistair Brooks, managing director of Commercial Real Estate Services Limited

Where a development has more than one occupier/tenant, most modern leases provide for common services costs to be shared via a service charge. When setting up a service charge 'scheme', these should be equitable, easy to understand and flexible enough to accommodate future developments.

Where all the occupiers are the same use-type (e.g. retailers trading nine to five, seven days per week), this is relatively simple: assuming similar lease provisions on cost recovery, one totals the costs and shares them using a common formula. Most modern leases provide for floor-area apportionment.

Trouble comes when there are varied occupier types (e.g. hotel and retail) as each will consume, and wish to contribute towards, services differently. There may also be an affordability issue.

How is the service charge scheme to be maintained?

We were asked to advise on revisions to a service charge scheme for a shopping centre, above which the owners were planning to construct a hotel. Had the existing floor-area-based charging scheme been applied, the hotel would have funded 75 per cent of the total service charge. Inequitable! A whole new scheme would have required amending existing retailers' leases. Unfeasible!

Liaising with the hotel, we listed all the services and then assessed which services they benefited from, and in what proportion. This produced an overall cost applicable to them, which amounted to a percentage of the whole. From that percentage, we calculated a 'notional floor area' to be used to calculate their future charges.

The issue, however, is not necessarily who the buyer is, but why they are seeking hotel investments. Broadly, the reasons are the traditional ones: capital growth and income – and hotels, as mentioned above, are rare in their ability to offer both, albeit the risk profile reflects such returns in relation to the investment.

However, to these generic, although appropriate, motives, we must add some specifics, some of which are unique to the industry. The most obvious are hotel companies themselves, who want control over one of the raw materials of their industry: the hotels themselves (their other primary raw material being people). Then we must consider the other end of the scale: the individual, whose motives can include ego (as one banker describes it, 'a variant of a boy's desire to own a pub and buy your friends drinks, just for slightly wealthier boys'), together with the escaping-the-rat-race lifestyle decision, although this is far more commonly associated with owning and operating, rather than investment.

Most investors are either looking for income or growth, or an acceptable combination of both. A variant, but an important one, especially given the rise in popularity of hotels in recent years, is the broad-based property investor.

Since the mid-1990s, hotel investment has become more and more attractive to companies who traditionally focused on commercial industrial and retail property. Again, the motive must be examined. In the early years, it was an entrepreneurial assessment of an undervalued asset class that merited exploitation. The likes of Land Securities, London and Regional Properties, Heron International, Starwood Capital and Blackstone Group all utilised their real-estate knowledge to create capital growth from hotel real estate while benefiting from higher yields. Mostly, they understood the increased risk associated with these investment decisions and the hotel formed part of a broad investment policy, with government-let offices at one end of the spectrum and luxury hotels at the other.

However, as yields have sharpened (and in this context, 'sharpened' means a lower yield or a higher price and, indeed, a longer period before the initial investment is paid back), some investors have entered the sector chasing yield.

The culture of leverage has also played its part. Traditional bankers used to say that the right amount of leverage was around 60 per cent by value, and two-times interest cover above earnings. Low interest rates, private equity, opportunity funds and, ultimately, the increasingly distant memory of the last recession all contributed to increasing leverage in all markets. In turn, this combined with the 'wall of capital', driving leverage levels in excess of 85 per cent in late 2007 for prime hotel investments.

Falling or sharpening property yields have the effect of making debt less serviceable, and yet the leverage issue can force debt levels up, not down, compounding the potential problems for hotel lenders.

This has resulted in investors seeking greater yield than has been available in their core real-estate sectors (typically commercial: offices and retail), which, consciously or not, can result in accepting higher levels of risk.

This has manifested itself in three main ways: the compression of secondary yields (and at a faster rate than primary property), the move into riskier

Understanding luxury hotel values

Graham Craggs, managing director – advisory, JLL Hotels and Hospitality

High prices have been paid for luxury or so-called trophy hotels, often in international gateway cities, that are sometimes difficult to rationalise with regard to the general tone of yield and pricing-per-room analysis prevalent in the market.

There are a number of reasons why this category of hotels may appeal to well-capitalised investors, who typically have a lower cost of capital than debt-driven buyers and owner-operators/brands. These include:

1. Long-term capital protection
2. Geographical diversification of capital
3. Prestige
4. Personal or intangible reasons
5. Underlying alternative-use value.

In this sector, at times, pricing has been driven higher by a shortage of opportunities, creating increased 'buyer tension' among an increasing pool of investors with new capital sources. But, yield aside, what has caused the extraordinary operating performance of some of these hotels that contributes to the overall sale price?

Trophy hotels are almost exclusively full-service in nature, with a wide product offer including a number of food and beverage (F&B) outlets that are charged at premium prices, as well as spas and extensive function space.

Critically, the room inventory inevitably includes a high proportion of suites, which are capable of being sold (dependent on their size) for many multiples of the standard room rate. The variance in terms of the mix of suites can lead to a material differential in performance from one hotel to another.

It is therefore critical that any analysis of pricing for this category of hotel fully accounts not only for underlying investor appetite for specific property attributes and locations, but also the operating potential that exists as a result of the individual characteristics of the hotel.

transaction structures, and movements into investment classes with greater yield. Primarily, these classes have been industrial and hotels. One result is a wave of new investors seeking hotels not for their underlying business, but for their combinations of yield and income.

The risk, and that underlying business that makes them what they are, is a by-product, but certainly one that some investors have found quite attractive once they were looking for new investments.

Yields

Yields tend to be counter-intuitive and can be quite confusing for the uninitiated. When talking about values, higher numbers seem as though they should be better, as it is logical that the higher the number, the more valuable it is. In this manner, logic dictates that a 20 per cent yield should be more valuable than a 10 per cent yield. Unfortunately, that is not the case in valuation terms, as the 'multiplier' applied to an income is the inverse of the yield. So, in fact, the higher the yield, the lower the multiplier, and therefore the lower the value.

A 20 per cent yield (at 1/20 per cent) gets a 5 × multiple, while a 10 per cent yield (1/10 per cent) gets a 10 × multiple, and as such is more valuable to an investor. In effect, the lower the yield, the higher the multiplier applied to the income stream.

As such, using the terminology 'higher' and 'lower' when talking about yields can be confusing. Instead, it is better to talk about 'sharper' yields (lower numbers) and 'softer' yields (higher numbers).

Types of hotels

As the hospitality world continues to expand, it appears as though there are more and more ways to categorise hotels. Remembering that each hotel is generally unique, however, it is quite possible to divide hotels under various different categories. This can be useful when comparing business models, trading records and values with other hotels. These categories would include:

- quality, for example, five-star hotels through to ungraded hotels;
- transport hub-type, for example, railway hotels or airport hotels;
- location, for example, city centre hotels or roadside motels;
- geographic region, for example, Parisian hotels or Auckland hotels;
- primary customer base, for example, tour operator hotels or conference hotels;
- size, for example, boutique hotels or large hotels;
- services provided, for example, full-service hotels or spa hotels;
- aparthotels, serviced apartments and long-stay properties;
- experience hotels, for example, unique properties or safari lodges.

Of course, it is never quite as simple as that, and one hotel could be a boutique, country house hotel with specialist spa and golf facilities, catering to leisure guests in the Black Forest in Germany. As such, it would have qualified to be categorised under the majority of the above.

That is the point about the uniqueness of each hotel – rarely do you find two hotels influenced by exactly the same market characteristics.

Quality

By their very nature, customers expect to pay differential rates for hotels, based on their perceived quality. A five-star hotel will almost always cost more to stay in than a limited-service hotel. Each level of quality should be meeting a specific market need and so should be no less successful (if well conceived in the first place), generating reasonable margins for the owner.

Transport hub hotels

This category of hotels includes railway station hotels, roadside lodges and motels, and airport hotels. Each have very separate operational issues that impact the way they do business. It has been said that the only hotels in the world not to experience 'a Sunday night' (traditionally a period of low demand) is an airport hotel. Railway station hotels tend to generate significant non-resident F&B spend, while roadside motels can have some of the highest conversion rate for sleeper-to-diner in the industry. Airport hotels, on the other hand, tend to have high demand for conference business, as well as demand from short-stay customers (four to six-hour stays), especially for 'airside' hotels.

Locational hotels

The differences between city centre hotels, suburban hotels, country hotels, seaside hotels, stadia-based hotels and principal town hotels are more varied than would first be imagined. The pressing need for car parking at a hotel, for example, is less in city centres where public transport and regular taxi services are available.

Country hotels need to provide a full F&B offering for customers, while stadia-based hotels can sometimes escape with providing a limited offering.

Locations also tend to dictate the level of demand and supply, which in turn dictates the services and quality needed to attract customers. Specialist hotels, such as stadia hotels, have an element of their market that cannot be taken away (the sports fan wanting to stay at the ground, or the visiting music fan wishing to be as close to the venue as possible), but even such hotels will need to attract more than one type of customer.

Geographic regions

This can include country-wide, city-wide or borough/state-wide, as is relatively self-evident. The point of such characterisation is that hotels in Santa Monica Beach will all, to a greater or lesser extent, be fighting for the same underlying business.

There will, of course, be some crossover in demand from other geographic areas, especially when demand is internationally or nationally driven (rather

Figure 1.4 An 'experience' hotel

than locally), but the key business drivers in a geographic location will remain important to all the hotels in the area.

Customer base

The simplest way of looking at the customer base is to divide it into business or leisure demand, although of course there are many others factors involved, and breaking down the demand base even further is useful. This is traditionally done by breaking leisure into further groupings such as 'group bookings' or 'fully independent travellers', and corporate customers into local, national or international, and again into conference bookings and group bookings.

Size

This might seem a strange division for hotels, given that it is just one small factor, but the size of the property has a major impact on the nature of the business done at that hotel. A smaller hotel typically has an initial disadvantage in terms of profitability. The lack of bedrooms can mean it has fewer economies of scale to cover the various undistributed and fixed costs associated with a hotel. For example, each hotel will require a general manager, but a 200-bedroom hotel has twenty times more rooms to cover that cost than a ten-bedroom hotel does.

Larger hotels usually have more buying power as well, so many operational costs are cheaper for them than for smaller hotels. However, on the downside, having too many rooms can lead to lower room rates to try and fill the higher number of rooms.

Figure 1.5 A 'boutique' hotel

A larger hotel sometimes has more difficulty providing a high level of personal service, as it can be difficult to remember the names of each of the 1,000 guests, compared to twenty guests potentially staying in the ten-bedroom hotel.

Services offered

The range of services offered by a hotel are a key differentiating feature, and one that is very important when looking at the trading profile, and therefore the value, of the hotel. Traditionally, a full hotel will offer a bed for a guest, along with F&B options, leisure facilities (e.g. a gym and possibly a swimming pool) and probably some meeting facilities. In contrast, a bed and breakfast hotel will provide just that: bed and breakfast. Limited-service hotels are somewhere between the two, usually providing a limited F&B option for hotel guests.

Aparthotels and serviced apartments

These could quite easily be included in the 'services offered' category, but because of the nature of these types of properties, they are considered worthy of their own category. These 'longer stay' options offer more of a home-away-from-home, typically offering the facilities of a suite or junior suite in a traditional hotel for the price of a room. Facilities can be as extensive as in a full-service hotel, though more commonly they are room-only, or limited-service at best. As such, location is important, needing to be a place where longer stays might be required, as well as being somewhere close to restaurants and bars.

All-suite hotels

All-suite hotels are different from serviced apartments, as they tend to be hotels first and foremost, rather than aparthotels or serviced apartments. All guest rental units consist of one or more bedrooms and may include a separate living area. Many suites contain kitchenettes or mini-refrigerators.

Experience hotels

Experience hotels are the type of property that a customer visits not just to stay the night as the primary motivation, but more to enjoy the experience of staying there. This would include some boutique hotels, themed hotels (the Beatles inspired hotel in Liverpool, Madonna Inn in California or Cinderella Castle in Disney World), specialised hotels (like ice hotels, lighthouses, tree houses or tented lodges) or hotels converted from another use (like the Four Seasons Istanbul, converted from the notorious prison in the movie *Midnight Express*).

Spa hotels are also included in this category, as the primary purpose of the stay is the experience rather than the destination itself.

Lifestyle hotels

Another category of hotels is the 'lifestyle property'. This can be a mix of the categories above but the defining feature is that it appeals to people who want to own the hotel or resort for the lifestyle it affords them. Typical properties that are commonly 'lifestyle hotels' include safari lodges, B&B establishments, boutique hotels, country house hotels and seaside hotels.

Complementary land uses – potential demand generators

The location of a hotel is of paramount importance to its success as a business. If the hotel is located in an area with strong demand and complementary uses, it has significant operational advantages. What constitutes a complementary land use will depend on the nature of the hotel. For example, a business hotel will benefit directly from being close to a business park or office complexes, if those surrounding businesses are likely to generate room-nights from their operation.

Certain types of 'industry' generate more room-nights than other types, so a four-star hotel located close to a pharmaceutical office complex is likely to generate more room-nights than if it were located close to a distribution centre.

Certain land uses are generally considered more complementary to hotels than others. These would include anything that attracts large numbers of people, including conference centres, music venues, sports stadia and theme parks. Other land uses can also be beneficial if handled correctly by the hotel, though

The challenges facing country house hotels

Peter Hancock FIH MI, chief executive – Pride of Britain Hotels

The country house hotel as we know it can claim a number of 'inventors'. These include Francis Coulson and Brian Sack (Sharrow Bay), Peter Herbert (Gravetye Manor) and Martin Skan (Chewton Glen), all of whom managed to create luxurious environments for their guests and offered them great food – a rare combination in British establishments at the time.

During the 1980s, the movement took off, fuelled by interest from American tourists looking to experience the English country house lifestyle they had seen in TV adaptations such as *Brideshead Revisited.* The organisation I work for, Pride of Britain Hotels, launched in 1982 as a small collection of independent properties happy to pool resources to tap into this growing market. Suddenly it was fashionable to brand one's hotel as being a 'country house' and the magic has lasted ever since, though it now relies less on overseas visitors thanks to a strong UK economy alongside frequent reminders of the provenance of our great houses provided by the National Trust, and yet more popular dramas including *Downton Abbey* and the latest version of *Upstairs Downstairs*.

I think it is no exaggeration to say that short breaks at lovely hotels in the British countryside have become part of middle-class living, as essential to our well-being as a dishwasher, membership of a gym or a quality car. So how can any hotelier fail under these benign conditions? I believe there are several issues that threaten to put a spanner in the works, though none of them is impossible to overcome.

1. A new breed of trendy hotels is flourishing. Examples include The Pig and the Soho House group but there are lots more. These places are in some ways the antithesis of traditional hospitality because they are more casual, less formal and anything but old-fashioned. Having said that, the best country house hotels have also reinvented themselves, attracting guests from a wide age spectrum thanks to their lovely spas, stylish décor and professional yet friendly staff.
2. Online travel agents (OTAs) and the growing use of electronic booking systems have created price transparency such as we've never known before, meaning that guests are often tempted to book through a commission-driven third party rather than direct with the hotel. This eats into profits and can lure all but the very best hotels into heavy discounting, putting their ability to reinvest in quality at risk. Again, there are reasons to be optimistic, as revenue management experts master the dark art of maximising achieved rates.

3. Staff costs remain a mighty burden on hotels at the top end of the industry simply because great service lies at the heart of what they do, and no matter how much technology you introduce, hospitality still needs lots of trained people to cope with the demands front of house, in the kitchen and in housekeeping. Falling unemployment and legislation are putting upward pressure on salaries, which must be passed on somehow. So if you can't be the cheapest, it's quite a good idea to be the best.

Long live country house hotels and the wonderful people who sustain them.

the advantage is not quite as obvious, for example, retail parks, nightclubs and cinemas.

However, there are also a wide number of land uses that have no direct impact on a hotel's trading potential, while some others have an adverse impact. Any use that would adversely impact on a hotels guest's experience is likely to deter guest use, and therefore impact negatively on value. Imagine a hotel located next to an abattoir, a nuclear (or indeed traditional) power plant or one located next to something very noisy or smelly. It is easy to see how anyone staying would be disappointed and, if they have already committed to stay, would chose not to stay again in the future. They would also be likely to put adverse reports on the Internet or to friends and family, which would deter other potential guests from booking.

Resorts tend to occupy larger land areas and are slightly less influenced by surrounding land uses, as there is usually distance between the guest accommodation and the surrounding use.

The location needs to be something that attracts customers – whether beach location, ski area, theme park or a natural attraction – and the facilities need to complement the whole stay of the customer. As such, there will usually be multiple food outlets and bars, as well as facilities to keep the customers occupied during the day, whether water sports or something similar.

What is a valuation?

It is amazing how often you hear the word 'valuation' mentioned and yet there is not one universally accepted definition of what a valuation is. According to the Oxford English Dictionary it is 'an estimation of the worth of something, especially one carried out by a professional valuer'. This seems to provide the general consensus of what a valuation is: an estimation of the monetary worth of something.

However, this provides a general basis for what a valuation is; it does not provide the full picture, as the value of something will depend on the purpose and type of valuation being carried out. There is, more often than not, more than one 'value'!

Why are valuations needed?

- Valuations are fundamental to real-estate transactions. Without knowing what the market value of a property actually is, it is impossible for the seller or the purchaser to know if they are agreeing the right price for a property.
- A bank cannot gauge whether to lend on the transaction, or how much to lend, without knowing the current market value.
- A developer needs to know that the hotel will be worth more than it costs to develop, and the only way to be certain of this is to commission a valuation.
- An investor, whether investing directly into the company as a principal or buying shares on a stock market, needs to know the value of the assets of the company they are investing in.
- A company needs to know what its assets are worth for future strategic decision making.

Types of valuation

There are a number of different reasons for carrying out a valuation, all of which may provide a different 'value'. This is quite normal, because the purpose of a valuation could be to provide an estimate of what something might generate if it were sold (market value), or what it would cost to rebuild (reinstatement value), and it is quite common for these not be the same. That does not make either 'valuation' incorrect.

It is therefore very important to ensure that the purpose of the valuation (and the methodology adopted) is clearly understood and corresponds with the actual requirements.

Market value

Market value is the price that the hotel would sell for if it were properly marketed. The Royal Institution of Chartered Surveyors (RICS) definition of market value is 'the estimated amount for which an asset or liability should exchange on the valuation date between a willing buyer and a willing seller in an arm's length transaction after proper marketing and where the parties had each acted knowledgeably, prudently and without compulsion'. Market value is probably the most important type of valuation for banks, developers and owners as it calculates the money that would be received if the property were sold.

When speaking with different potential buyers, they will generally discuss what a hotel is worth to them. Perhaps one hotel owner will say that they 'really need a hotel in Accra' (as it is a vital but missing location for their business), or that they 'already have too many hotels in New York' and therefore do not really want another – such factors will affect the price they would be prepared to pay for the hotel. All these individual 'calculations of worth' will be reviewed by the valuer when assessing market value.

Reinstatement value

Reinstatement value calculates the cost of rebuilding the hotel in its current condition. This is a very important type of valuation and is the basis for insurance cover. It is important to note that this type of valuation is rarely the same as the market value as it deals with the cost of rebuilding the hotel, rather than what the hotel would sell for, if it were placed on the market. In strong trading areas, the rebuilding cost is likely to be lower than the market value, but if the hotel is located in a poor area where trading is weak, the rebuilding value may be higher than the market value.

Book value

Book value is a notional assessment of the hotel for accountancy purposes. It is influenced by the purchase price/development cost and the depreciation and other accounting policies of the owning company. Although the book value may have an impact on the price that the owners would consider selling for, it is actually more an 'estimate of worth', as it is a highly individual valuation and does not reflect its value to potential purchasers.

There is a significant difference between worth and value, mainly down to what is being calculated. A hotel might be worth US$100m to one operator, as their cost of capital is such, and their earnings potential is such that they could afford to pay US$100m for the asset. However, if they were the only party able to earn those sort of returns from the property, then this calculation would be an estimate of worth because of the individual nature of the assessment. However, if a number of other people were able to earn those sort of returns, and were prepared to also offer US$100m, then this calculation could become a 'value', as it is representative of the wider market.

Summary

A valuation is important for many different reasons but the specifics of each instruction need to be reviewed to check they are suitable for the individual needs of that type of valuation, to ensure the correct 'valuation' is being undertaken.

It is not unusual for a valuation for one purpose to be different from the valuation for another purpose. Indeed, a hotel could be worth more to one person than it would be to another, and yet neither of these 'values' may actually reflect market value.

It is vital that the valuation is undertaken by someone competent, professionally qualified and experienced in the relevant hotel market, and that the valuation report is read in detail, and looked at in conjunction with the trading projections.

2 Ownership guide

Types of ownership

Hotels, resorts and their associated leisure properties can be held in a variety of manners (ignoring tenure differences). In simple terms, there are three different categories of hotel ownership:

1 Vacant possession without encumbrance; that is, owned and operated by the same party.
2 Owned by one party but operated by a second party as a tenant. In this case, the operator has a legal leasehold interest in the property.
3 Owned by one party but operated by a second party as a manager. In this case, the operator is employed by the owner to manage the property on their behalf and they have no legal interest in the property itself, although they do have contractual rights to operate the property.

To complicate matters slightly, all properties that are not owned and operated by the same party (all but vacant possession properties) are referred to as 'investment' properties, despite the fact that most vacant possession hotels still require a substantial injection of money and so also represent a significant 'investment'.

Then in addition, there is a further complication in that the property could be franchised, either by the owner/operator, a tenant or a management company:

- Owned and subject to a franchise agreement.

Although both investment and vacant possession properties can be franchised, it is generally owner-operators (vacant possession) who see the most benefit of such a franchise agreement more than independent (non-branded) professional management companies.

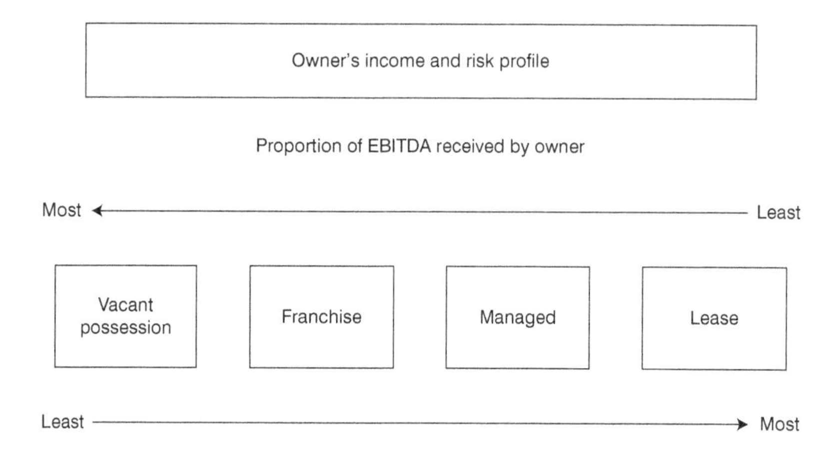

Figure 2.1 Owner's income and risk profile

Vacant possession

Vacant possession in the context of hotels is different from most other types of property, for example, retail or offices. If an office is sold with vacant possession, it means that the property is empty, that is, without a tenant or occupier of any type, and therefore without any immediate income. As such, the value is typically lower with vacant possession, as the owner will usually have a void period before any income is received.

A hotel is, however, sold as an ongoing business, and so to sell it 'empty' would adversely affect the trading. Vacant possession in relation to hotels means 'without a tenant or manager in place' (i.e. without any legal operational encumbrances). It will still be operational at the time of purchase and the purchaser will benefit from future bookings. This also means that the purchaser will be taking over the staffing liabilities of the property, and will also take over the stock (although this will be an additional sum to the agreed purchase price, calculated on the day of the transfer, at cost).

Globally, most hotels are purchased with vacant possession, with the owner also operating the property. The level of vacant possession hotels varies in each country and initially, as a rule of thumb, the countries with the greatest number of internationally branded hotels tend to have the lowest percentage of vacant possession hotels.

However, once the hospitality market in a country becomes mature, the level of franchising tends to increase, resulting in a swing back, leading to an increase in vacant possession properties, albeit subject to franchise agreements.

Stock audits

When a hotel is valued, the stock levels are not included in the valuation. The same is true when a hotel is sold: a separate stock audit will be undertaken and a price will be paid (at cost) for any stock that is transferred with the hotel. Many times it has been suggested that stock should be included within the agreed sale price, but this tends to be a mistake for all parties.

A hotel is sold as a trading entity, and comes with the benefits of future bookings in place. If the seller knows they are disposing of the hotel and decides to run the stock down so they can take more profit out of the business during the due diligence process, it can have an adverse impact on customer satisfaction and therefore on future bookings. Running out of a popular item (by not having sufficient stock) can also lead to lost sales during the due diligence process, thereby hurting the seller.

As such, it is usual to employ a third-party stock auditor to assess stock levels, for which a fair price (cost) will be paid. This is in the interest of all parties, even if it does add slightly to the due diligence costs of the transfer.

Owned by one party and operated by a tenant

This is the simplest type of investment. The owner creates a lease allowing exclusive use of the hotel by the tenant. This type of arrangement is common in commercial property investments around the world, and institutional investors are comfortable with the basic concept of a landlord and a tenant relationship, where responsibilities and rights are clearly defined.

A surprising large proportion of hotel investments across Europe are subject to operational leases. The owner of the property will either accept a premium at the start of the lease, collect rent based on the agreement struck between the parties (this will be affected by the terms of the lease) or some combination of premium and rent.

The actual structure of the deal will generally depend on the needs and intentions of the parties. The tenant will then be responsible for the operation of the property and will employ all the hotel staff themselves. They will undertake whatever repairs are specified within the lease at their own expense, and they will be responsible for all things that are specified within the lease as the tenant's responsibilities.

Owned by one party and operated by a manager

Another type of 'investment' is the 'managed' hotel. This is probably the most important single section of the hotel investment market in terms of hotel numbers.

Figure 2.2 Vacant possession hotel

The majority of hotel investments involving US branded hotels comprise hotel investments being run by a hotel operator on a management contract.

In effect, the owner is paying a manager to come in and run the hotel on their behalf. The staff, the repairs, the operational costs are all the responsibility of the owner, while the manager looks after the day-to-day running of the hotel.

The manager will charge a fee for their services, usually a management fee (based on revenue) and an incentive fee (based upon profitability). In addition, there may be charges for being part of a CRS (central reservation system) and/or booking charges.

Operated under a franchise agreement

A franchise agreement is a contract that allows a hotelier/hotel owner to use the name (and usually have access to the CRS) of a brand in conjunction with the hotel, whether Marriott, Hilton, Ibis or Park Inn, etc. There will usually be minimum standards that have to be complied with before a brand will allow the hotel to be 'branded', for example, room sizes, facilities, life safety and service level. The hotel will normally pay a fee based on turnover, rooms' revenue or booking numbers.

In return, the hotel will be branded (hopefully with a brand that actually helps them generate more revenue than it would otherwise generate), and it should benefit from central marketing and bookings.

A hotel can be owned by one party who employs a manager (paying a management fee) and is subject to a franchise agreement (also paying a franchise fee).

Figure 2.3 Managed hotel

An independent operator will have to persuade the hotel company of their ability to comply with the requirements of the brand, as well as their general good character and management experience and competence, before they will be able to negotiate terms for a franchise agreement.

Legal agreements and how they are structured

Leases

There are many different types of lease that hotel and resorts are operated on: from ground leases, shell and core property leases, through to fully fitted leases. Each type of lease has its own peculiarities.

A lease divides the legal interest in a property between a landlord (the owner) and a tenant, with different legal rights and obligations. The nature of the lease and the terms contained therein will dictate the various rights of each party, as well as their obligations.

Ground leases

The simplest and longest running type of hotel investment is a ground lease. Typically, the landlord will collect a rent from the operator who will have invested their money into building the property. The lease will be for a long period, to make the investment in constructing a property on the land viable. Fifty-year terms tend to be the minimum term acceptable, with 99- and 125-year terms quite common. Sometimes the term will extend to as long as 999 years.

The value of brands

Graham Dodd – Hilton Worldwide

The flexibility associated with property assets that are free and clear of any encumbrance has long been the holy grail, something to which many investors still aspire to this day. To some degree that still holds true, but the exponential growth in mobile and digital means that hotels as a real estate asset class present very different challenges to the mainstream.

'Free and clear', but at what cost?

Cashflow, profitability, the requirement to remain competitive and regularly reinvest in the physical fabric of the building are critical to their successful operation, their ability to raise finance and their liquidity in the market.

Hotel valuers providing advice to clients must have a thorough understanding of both real estate and operational issues, and to be able to assess the quality of the incumbent hotel management.

Over-dependence on OTAs to support a hotel can be very expensive, so how do hotels attract, win and retain business directly rather than through intermediaries such as OTAs?

Global hotel brands are at the forefront of this customer engagement and their brands' portfolios are growing dramatically. Driving more business direct to hotels is absolutely critical through global distribution systems, multilingual web sites, killer apps, mobile-friendly e-marketing and, perhaps most importantly, recognising and properly rewarding guest loyalty.

Access to demanding and expectant customers in the digital world in which we live is critical to the success of all businesses, not just hotels. These demands and expectations will only increase in the future.

Making the right choice about confronting this issue can be extremely rewarding but failure to address this challenge and remain competitive can be terminal.

The rental stream can vary: from a peppercorn rent (a notional rental amount that is not usually demanded or paid), to a fixed annual sum throughout the term, a rent that is reviewed periodically – based upon the rental value of the site, or a rent linked to a proportion of turnover or operating profit.

It has latterly become common for good sites to be leased to operators on a turnover basis. This ensures that the landlord benefits from the investment, while at the same time not unduly penalising the operator if trade declines. This flexibility usually results, over the term of the lease, in the owner of the land receiving a higher rent. It provides the operational flexibility required by the

Peppercorn rents

A number of ground leases specify that the annual rent is to be 'one peppercorn, if demanded'. For a contract to be valid, one party needs to pay something to the other party and it has become the convention for contracts where a rental payment is not actually required by the landlord that this 'consideration' is in the form of 'a peppercorn'.

However, the first mention of a peppercorn in a lease was in the sixteenth century in the UK, and at that time peppercorns were very valuable commodities, having to be transported by sea over very long distances. At that time, a peppercorn rent would have been extremely valuable to the landlord and a great imposition to the tenant, rather than having the nominal value that they possess today.

tenant to ensure that they can ride out any difficult markets. As such, it is popular with both landlords and tenants.

Typically, ground rents are considered very secure as it is unlikely that the tenant would risk losing 'the building' by defaulting on a relatively small ground rent payment. However, the actual income stream from a turnover ground lease is uncertain and therefore not viewed as quite so secure and, as such, the yield tends to be 'softer' than for other ground lease investment types.

The key terms for a ground lease are as follows.

The term

The length of the lease will impact upon the value of the lease to both parties. If it is too short, the operator will have a shorter period in which to recoup the building costs and so will not be able to pay as high a rent. If it is too long, the investor's residual value (when they can regain possession of the land) will be low, and therefore a higher rent (or initial premium) will need to be paid.

The demise

The extent of the land or building(s) that is included within the lease will impact on the value of the lease. If the land extends to a significantly larger plot than the hotel building actually occupies, it may be that at some stage in the future this land will have additional value; as such, the tenant may be prepared to pay a higher rent. Conversely, if the land included within the demise is too small (for example, there is no room to expand the hotel or provide adequate car parking) then a lower rent (or initial premium) is likely to be paid. If the demise includes any buildings, services, roads, parking or other beneficial facilities, it may enhance the rent paid.

Payment of the rent

The rent, and any mechanism for review of that rent, will generally reflect the level of any initial premium paid for the site. If a large sum has been paid to secure the site (known as a premium), then minimal rents may be charged. It is not uncommon for the landlord to lease the site rather than sell it, despite not getting a significant annual rent, just to keep control of the use of the site if it adjoins other land they still control.

A large number of ground leases specifically control the use of the site (and any building on them), enabling the landlord to have effective control of the use of the site throughout the term of the lease. A lease without 'use limitations' is more valuable to the tenant.

Where the rent is based upon a percentage of turnover (typical at airports, for example), it is usual to see clauses that enable the landlord to exercise some control over this income, whether by allowing them to review and comment upon the accounts, keep-open clauses or to enforce repairing obligations upon the tenant.

Repairing obligations

Where the landlord is keen to control the ongoing use of the site as part of a larger 'estate management' programme, fixed repairing requirements may be set.

Needless to say, each ground lease is different and certain leases will have contrasting 'key clauses'. It is important that the investor reviews all the clauses to see what impact they may have on value.

Traditional occupational leases

A traditional occupational lease is typical of the property market as a whole; the hotel building will be leased to the operator. Typically, the term will be shorter and the contract will contain more restrictions and obligations than would be placed on the tenant in a ground lease.

Some of the key terms are similar to a ground lease, others can be quite different. They include the following:

Term

Around the world, the average length of the lease can vary, typically between one to eighty-four years, although shorter terms seem to be becoming more common place in recent years.

Break clauses are sometimes included within the terms of the lease, allowing one or either party to the lease to terminate the lease early. The desirability of the lease length, to both the landlord and the tenant, will depend on the relative strengths of the market: in some markets, tenants will be prepared to pay a premium for a longer term, whereas, in other market conditions, break clauses may be more valuable to tenants.

Demise

The demise will determine what is included within the lease, as well as whether the building is fully fitted or built to a shell and core finish. In older leases, it is quite usual for hotel leases to be let on a shell and core basis, allowing the operator to fit-out the hotel according to their own specific operational requirements.

Rent reviews

How often the rent is reviewed, indeed if reviewed at all, will often have an impact upon the value of the investment for both the landlord and the tenant. The methodology behind the review will also be important. The additional certainty provided by RPI-linked reviews is prized by certain investors, above potentially higher (if slightly riskier) uplifts associated with review of the market rent, or those relating to a proportion of capital value.

Rent (and how it is calculated)

There are many different ways to structure the rent, from turnover leases, to part-fixed part-EBITDA-based rents, through to fixed rents which are annually increased. The type of rent, and level of the agreed rent in relation to market norms and affordability, will affect the desirability of the investment.

Alienation

Alienation is the transfer of a legal interest from one party to another. It might be a transfer of the whole of the property interest, or part of the property interest.

One of the key investment criteria for an investor is the quality of the tenant. As such, it is very important that the clauses that allow for the tenant to change are reviewed and 'priced accordingly'. Although absolute prohibitions on the tenant assigning their interest can be considered onerous to the tenant (potentially impacting on the length of lease they will consider and the rent they are prepared to pay), it is important to allow qualified prohibitions, to ensure that the quality of the next operator can be authorised by the owner, thereby ensuring that the investment value of the property does not decline.

Specific alienation provisions for hotels may include: the proposed assignee operating more than a specific number of hotels or bedrooms (to ensure the experience of the assignee); requiring specific market experience appropriate to the hotel in question; demanding the financial standing of the proposed assignee is equal or better than the existing tenant; and insisting that the landlord's interest is not adversely affected by the assignment.

In the case of turnover leases, there is often a blanket prohibition on sub-letting as this can adversely affect the landlord's income, as shown in the following example.

An unexpected problem with turnover leases

Turnover leases are seen as a fair way for landlord and tenants to share the operational risk of the hotel, where trading is not established or secure. When trading is good, the landlord benefits by receiving a higher rent, but if the trading declines, the tenant has a lower rental obligation, allowing them to continue trading without fear that the rental obligations will unduly harm the business.

However, there can be a largely unforeseen problem with turnover leases. Let us assume that a hotel with 400 bedrooms pays 25 per cent of turnover by way of rent, which could be within market norms. In this hotel, there is a fine-dining restaurant that runs at a profit margin of 15 per cent and a quality spa that generates 20 per cent profit.

In this case, the tenant loses money for every dollar that is generated in either of these two facilities, and it would therefore be in the tenant's interest to close both operations down. It is very unlikely that it was the intention of either party when the lease was agreed to have both parties working at cross purposes. The essence of a turnover lease is that it is a partnership – in the good times, the operator can pay the landlord more rent and in the poor times, the landlord is prepared to accept less rent to help the tenant to work through the lean times.

As such, it could be considered sensible to have different turnover percentages for the various parts of the business, assuming the revenue allocation is clearly defined. It would be a shame if 20 per cent of rooms' revenue is paid as rent and 15 per cent of F&B revenue, only to discover that a bed and breakfast rate offered by the hotel of $100 was internally allocated as $90 for the breakfast and $10 for the room.

EXAMPLE: HOTEL RICHARDS

The Hotel Richards has been let for ten years to an operator paying 20 per cent of turnover as rent. The operator earns €500,000 revenue from a restaurant/bar in the hotel each year, of which €125,000 converts to operating profit (25 per cent).

However, a third-party tenant is prepared to pay €200,000 rent for the property, and such an agreement would be in the interests of the tenant, as the income received by them would be greater.

However, as the landlord is receiving 20 per cent of turnover, if the operator runs the restaurant themselves, then the landlord receives €100,000 rent from this part of the building. In the event it is sublet, the landlord will only receive €40,000.

Repairing obligations

The allocation of responsibility for repairs will impact on the value of the investment. Full repairing and insuring leases (FRI), where the tenant is responsible for the repairs to both the exterior and the interior of the property, are still the standard within the hotel industry. However, more and more internal repairing leases (IRL) are being seen. This is where the tenant is only responsible for the interior of the building and the landlord is responsible for the external structure. In addition, various hybrids of the FRI and IRL are being created, where part of the outside structure is maintained by the owner and the remaining repairing liability remains with the tenants.

Certain leases go as far as specifying certain minimum levels of repairing expenditure each year (for example, 7 per cent of turnover will be spent on repairing the property). This is one way to ensure the upkeep of the property from the owner's perspective, but it can be quite onerous to the tenant, especially if the level of expenditure is higher than would normally be required.

Insurance

It is usual for the tenant to bear the responsibility for the cost of insurance, although the landlord will be keen to ensure that they have a mechanism to insist that adequate insurance is in place. It can be more expensive for a hotel company to have the insurance purchased by the landlord (and for them to reimburse the landlord), because some of the larger hotel companies are able to command significant cost discounts by placing a block insurance policy across their whole portfolio, rather than on an individual-by-individual property basis.

Alterations

A blanket restriction on alterations can be quite prohibitive and will potentially affect the rent that a tenant will be prepared to pay, especially if the term of the lease is longer than fifteen years. The hotel industry requires flexibility to be able to ensure that a property can be altered to match the needs of an evolving customer base.

Keep-open clauses

These are rarely used in hotel leases, although they are sometimes used in conjunction with turnover leases. By their nature, most hotels need to remain open to trade and so, unless the business is seasonal, it is not considered unduly onerous for a keep-open clause to be included in the lease terms.

User clauses

These can be quite simple, for example 'to use as a first-class hotel'. They can also be inserted to have an impact upon the potential trading of the property, for example, 'not to sell alcohol', 'not to use for gambling' or 'not to hold auctions'. They can also restrict the property from being used for other, potentially more profitable, uses.

Optimise/maximise the turnover/EBITDA

In most leases where turnover or EBITDA are directly linked to the rent payable, then there is often an obligation on the tenant to maximise (or optimise) trade. The difference between maximise and optimise is quite important.

'Maximise turnover' means to generate the highest amount of turnover possible.

'Optimise turnover' on the other hand balances the long-term future of the property with maximum turnover generation, ensuring a short-term advantage is not exploited to the detriment of the hotel over the longer term.

Asset management rights for the landlord

In most performance-based leases there is provision for the landlord to review the basic marketing plans and budget produced by the tenant, to ensure that the company is doing its best to maximise (or optimise) turnover and EBITDA (and therefore paying as much rent as possible). If the landlord does not have these rights, it can adversely impact on the long-term value of the investment.

Access to operational management accounts

Ideally, the lease should provide access for the landlord to review the management accounts for the hotel. When the investment is being valued or sold, having access to the operational accounts will be beneficial to the investor, who will then have some certainty regarding the affordability of the rent and the vacant possession value of the unit.

Forfeiture

Clauses that are considered unduly onerous, and which impact on lending institutions' security over the asset, will have an adverse impact on a landlord's value.

Mandatory FF&E expenditure

Some leases detail specific provisions for the annual expenditure of FF&E. Needless to say, if this is set at the right level, it will be attractive to the landlord as it will ensure that the property is kept in good condition while not being a disadvantage to the tenant. However, if the FF&E spend is set too high, it can be a hindrance to the tenant, who will not be able to offer as high a rent as they otherwise would have done (and indeed it will deter some potential tenants), thereby affecting the value of the landlord's investment.

Operating leases with fixed rents

Hotels are often let out to tenants on the basis of a fixed rent. In such leases, the demise may be a fully fitted hotel, a 'shell hotel' with no fittings and, indeed, almost anywhere in-between these two extremes.

The variety of this type of lease is almost infinite. The term can vary from one year to over fifty years, the repairing obligations can rest solely with either party and, again, almost every lease covenant can be varied.

In terms of rent, it is common to see fixed rents granted as follows:

- a rent specified for each year throughout the whole term of the lease;
- fixed rents annually indexed, for example, reviewed in line with RPI or CPI;
- fixed rents with reviews every five years linked to RPI or CPI;
- fixed rents linked to the market rent (of hotels and/or other classes of property);
- fixed rents linked to a proportion of market rent;
- fixed rents with review to a proportion of the capital value of the hotel.

A lease with a known rental stream throughout the whole term is considered more secure by investors (assuming identical levels of rent, lease terms, quality of tenant and all the other lease covenants) than one that is linked to market rent.

Annual reviews are also considered more desirable (and therefore produce prima facie sharper returns) because there is a perceived lower level of risk than those that are reviewed less frequently.

Operating leases with performance-based leases

There are also varying categories of performance-based leases, including: those that pay a proportion of turnover, those that pay a proportion of the hotel's net profit (EBITDA) to the landlord, and those that have an element of guaranteed income and an additional element based upon an agreed percentage of turnover/EBITDA where it exceeds that guaranteed element of rent.

Once again, the more secure and the easier it is to collect the rent, the sharper the returns are likely to be. EBITDA-based leases, where the landlord is reliant upon the successful trading of the hotel, transfer a high proportion of the 'operational risk' across to the investor, and, as such, the investor is likely to demand a higher return to compensate for this additional risk.

Management contracts

A management contract (or hotel management agreement) is similar to a turnover lease but has the advantage of avoiding a balance-sheet liability for the operator, while providing potentially easier provisions for re-entry for an owner when the operator is not performing.

However, there are three key investment differences between a turnover lease and a management contract:

1. employment liability risks
2. repairing liability risks
3. operational risks.

A management contract is an unusual structure for many types of investors as it transfers most of the 'operational risk' across to the investor, at the same time as transferring the property risks too. The investor behind a management contract will be employing the operator to manage the property on their behalf. This means that the staff are all employees of the investor, along with all the relevant employment regulations and legislative issues this raises. In addition, the repairing liability for the property rests squarely on the shoulders of the investor. The operational risk is also held by the owner, and if the manager (or management company) is not very good then the investor's returns will be adversely affected.

The first thing to say about a management contract is that it should always be viewed as a partnership between the owner and the operator, rather than in any other way. If one side has too much control of the partnership (the terms are too favourable to one party for it to be a fair partnership) then it tends to fail as a relationship.

The investor needs to earn a reasonable return on their investment, while operators need to generate reasonable management fees. If either of these 'returns' is too low, one party will be dissatisfied with the agreement and will have no reason to ensure it works effectively.

Some of the key terms in a management contract are as follows.

The 'term'

Most terms tend to be for between ten and twenty-five years, although some agreements provide that, at the operator's option, the relationship will last for over fifty years. Any period outside the usual range could have an impact upon the value of the investment.

If the management contract 'adds value' to the property, then, from the investor's perspective, the longer the term the better. If the property is highly desirable and likely to attract multiple potential operators, then a shorter term may add value to the underlying real estate interest.

Whether any extensions to the term are mutual or at the discretion of the operator can also have an impact on the value of the investment.

The level and detail of management fees

The cost of the management of the hotel by the operator will impact upon the cashflow generated by the owner, and as such will have an impact upon the value of the investment. It is usual to see a base fee (which is a percentage of turnover) and an incentive fee (usually a percentage of 'adjusted gross operating profit' (AGOP)).

The level of fees will vary depending upon the locality but fees in the region of 2 per cent to 3 per cent of turnover as a base fee and 6 per cent to 10 per cent of AGOP as an incentive fee are typical. AGOP can be calculated in many ways, depending on the contract. It is advantageous for the investor to deduct as many 'fixed costs' as possible before calculating the incentive fee, if that can be agreed.

Typical adjustments to gross operating profit (GOP) (to arrive at AGOP) might deduct the base management fees, the FF&E reserve, property insurance, property taxes or any combination of the above.

Certain contracts allow for incentive fees to be subordinated to debt service or some other agreed performance criteria. Subordinated fees may be waived or accrued, depending on the relative strength of the negotiating parties when the agreement was being drawn up.

In certain circumstances, unpaid fees become debts chargeable upon the property, which, needless to say, has an impact upon the value of the investment.

The operator system fees

The way that 'other charges' are handled will affect the income generated by the investment, and therefore its value. Reservation fees can be charged per booking or per room-night, or they can be a percentage of revenue generated or a fixed price per booking. They can be charged at the time of booking or when the guest arrives, and sometimes reservation fees accrue in the event of no-shows. All of these factors can have an impact on the value of the interest to the investor.

Shared services

It is usual for the owner to have approval of those services which are to be shared with other hotels operated by the operator. These services should be detailed in the agreement. If there is no control over such shared facilities (for example, the auditor for a competitive hotel run by the same management company has the right to review the hotel accounts), it can have a detrimental impact upon the value of the investment.

Performance test criteria

Many management contracts contain no performance test criteria at all, although the better-drafted contracts (from an investor's perspective) provide the right for the owner to terminate the contract, without compensation, should the operator consistently fall short of the performance test.

A performance test will usually specify a target that the management company would have to attain if it were not to fall foul of such a provision. For example, the management company may be tasked with achieving 90 per cent of the agreed budget (possibly in terms of revenue and GOP), or achieving 90 per cent of the revenue per available room (RevPAR) of an agreed competitive set of comparable properties. It is usual that the performance test requires repeated failure, rather than a one-off 'underperformance', and typically this is for three consecutive periods.

Choosing the competitive set[1] for benchmarking purposes is quite subjective and these can sometimes be unsuitable. Any defect in such a competitive set can adversely affect the value of this clause to the investor.

Owner's right to sell the property

The owner should normally be able to sell the hotel without the operator's consent, although many older agreements have a blanket restriction on the owner selling the hotel without the operator's approval. If a purchaser wishes to change management companies, or is itself a management company, there is usually a right to terminate the management agreement, with compensation paid to the operating company.

In some management contracts, the operator has a right of pre-emption and can take over the deal after it has been negotiated. This can have a dramatic impact on the value of the investment as it can substantially lower the number of investors prepared to go through the due diligence process. In this instance, such investors know that at the end of the process it may have all been in vain, despite the truest intentions of the vendor, and they could be left with the very real expenses they have incurred through the process.

Owner to approve budgets

Most management contracts state that the operator will provide the owner with detailed operating budgets, a breakdown of proposed capital expenditure, and a marketing plan for the owner's approval, each year. If the owner does not have this right, it could adversely affect the owner's investment.

Owner to have the right to approve the general manager and financial controller

Most contracts will give the owner the right to approve the appointment of the property's general manager and sometimes other key personnel like the financial controller. Without the ability to question the appointment of the general manager, the owner is in a weaker position to ensure the optimum performance of the investment.

FF&E replacement reserves

These should be appropriate to the quality of the asset, whether set at 3 per cent of turnover per annum, 4 per cent, 5 per cent or at even higher levels of turnover. It is in the owner's interest to have this clearly specified, because if an adequate

FF&E reserve is not deducted, such expenditure will become a capital expense in the future. Although there may be tax advantages to such a course of action, it could result in higher incentive fees being paid to the operator.

Any shortfall in FF&E expenditure should be carried over to the next year, and should not be credited to the GOP (otherwise additional management fees may be paid on such an under-spend).

Redevelopment in the event the property is destroyed

Some operators request a clause that obliges the investor to rebuild the property in the event it is destroyed. The owner should have the freedom to choose not to rebuild the hotel in this instance, as having the freedom can sometimes lead to a potentially higher alternative use and therefore value.

Restrictions on competition

The contract may limit the number of hotels that the management company can operate within a 'competitive' geographic locality to the subject hotel. It may be that the restriction is merely on the use of the brand name of the unit.

The greater the restrictions, the more control the investor has over the future openings operated by their partner affecting their investment. The logic is that the investor has agreed to have a 'Holiday Inn' brand which they feel is beneficial to the trading potential of their hotel. If another 'Holiday Inn' opens very close to the property then some, or all, of that benefit is lost, and the reason for entering into the agreement on those terms has been lost.

Alternatively, or in addition, the owner may have a priority listing in the operator's reservation system (i.e. always first choice) or the right to approve any additional use of the operator's name.

Bank accounts to be the property of the owner

The bank accounts are usually held in the name of the owner and not 'held in trust' by the operator. The owner will usually also have the right to approve the signatories to the account. If they do not have these rights, it can be detrimental to the value of the investment.

Branded operator's equipment

This should be kept to a minimum to avoid any additional expense on behalf of the investor on termination of the management contract. If the level of branded equipment is unusually high, then it could have a detrimental effect on the residual value of the investment.

Arbitration

Most contracts allow for arbitration in the event of disputes between the parties. However, where the chosen seat of such arbitration is, and the applicable law, can have quite an impact on the value of the investment.

Independent external auditors

The investor should have the right to appoint their own accountants and auditors. If they do not have this right, potential investors may be deterred from taking on the investment for fear of their advisors not being seen to be sufficiently independent from the operator, thereby affecting the value.

Asset management

What is asset management, in hotel terms?

Assuming that the hotel is not being operated by the owner, it usually proves prudent to employ an 'asset manager' to liaise with the operating company. This is to ensure the owner's interests are uppermost in the manager's thoughts. Many investors are not hotel experts and indeed may have no specialist knowledge regarding hotels or hotel investments. The asset manager will review performance, budgets, marketing plans, maintenance schedules and capital investment programmes to ensure everything is being done to enhance the owner's returns.

The asset manager should be looking to answer the question, 'Is it the right product in the right place?' There is strong demand for luxury resorts in certain locations, while in other areas mid-market business or budget hotels are the optimum type of hotel operation. If market demand changes, can the hotel's position be realigned?

Targeting the hotel to the most appropriate segment (including quality, size and ancillary facilities) is one of the most important steps to ensure the owner can maximise their investment returns.

However long the investor intends to own the hotel, the asset manager should ensure that trading is optimised. The asset manager's responsibility is to ensure that the long-term value of the asset is not sacrificed to short-term income, without the owner being aware of that risk.

It is important not to neglect basic repairs and maintenance. If the property is not kept in good repair, it has an impact on both trading and capital value, and that impact is always greater than the short-term savings generated by under-investing in the fabric of the hotel.

Treasure hunting

The asset manager will also be involved in a thorough 'treasure hunting' exercise. Treasure hunting means looking at the hotel or resort and establishing if there are

Management contracts should be a partnership

Richard Bursby, partner – Taylor Wessing LLP

Hotel management agreements (HMAs) are unusual contracts. Not only are they long-term contracts (typically fifteen to twenty years), but they relate to the management of a business by an operator on behalf of the owner who owns that business. HMAs are therefore 'relational' agreements; this is not a one-off supply of services, but an ongoing provision of services in a long-term relationship. They also need to be fluid – to cater for differing circumstances and events over time. Like all successful long-term relationships, there need to be parameters and compromise as well as proper incentivisation to drive good financial performance. A contract where the operator has total control means they will lose sight of the fact they are running a business in the best interests of the third-party owner. Nor is it positive where the owner has excessive controls, preventing the operator doing what they have been hired to do – run the hotel – and where the operator is not properly incentivised to drive profitability at the hotel.

The successful HMA gives certainty of term and defines the parameters within which the operator is free (and motivated) to do what they do best – operate the hotel – and outside of which it needs the owner's control. Overlaying this are provisions allowing the owner to have full visibility over its business, and a degree of control over the operator and their performance.

An unbalanced HMA will not stand the test of time. A good example is the Cadogan Hotel which was for a short period run by the Stein Group under an excessively owner-friendly HMA by which the operator could be evicted on five minutes' notice with minimal compensation. It is a case study in disincentivisation for an operator to build goodwill. For example, any operator, when faced with the choice of which general manager to place in a hotel – perhaps the single most critical decision to be made – will not choose the best candidate if there is no reward to building a long-term business.

any unexploited areas where performance could be enhanced to add value to the operation of the property, as well as to the capital value of the real estate.

There may be the possibility of adding rooms to the hotel, assuming such rooms are needed. Alternatively, there may be space for a spa, and having a spa may enhance bedroom demand during the quiet weekends, or attract more 'suite users' to the hotel, thereby enhancing the average daily rate (ADR) at the hotel. It could be that changing some of the bedrooms to meeting rooms will enhance overall trade, or that developing private residences on the golf course will provide valuable, unexpected revenue.

Anything that adds profit to the property will enhance the investor's income, as well as strengthening the value of the property.

Operator selection

Trevor Ward, managing director – W Hospitality Group

At the beginning of 2015, the international and regional chains had signed deals to brand 270 new hotels with almost 50,000 rooms in 40 countries in Africa. Of the top five countries by number of rooms, four were from North Africa (Morocco, Algeria, Tunisia and Egypt), with Nigeria in first place with almost 8,600 rooms.

Why engage an operator, when so many hotels do without?

One of the first reasons, intertwined with several others, is that you might not be able to get finance without one. Several banks, local and international, will not lend to a hotel project unless they know that professionals are in charge. Why? Because that reduces the risk of default on the part of the owner, with experts responsible for and incentivised to generate profits for debt service. And those profits should, all things being equal, be higher than if the hotel was unbranded and run by the owner because guests will pay more for a branded product than one without a brand (test yourself on that in the supermarket one day!).

Brands bring proven training programmes and the opportunity for secondments to their other hotels. They bring stringent financial controls, menu concepts, bespoke software for maintenance and other management, and experience of how to solve problems encountered in the past.

And for all this, they charge fees.

Not every hotel needs a branded operator. Many hotels benefit considerably from them, particularly in a competitive scenario. Whether to brand or not to brand: take time to investigate the options – remember that a management agreement runs for a longer period than most marriages survive!

Selection of operator

Assuming the investor is not planning on operating the property themselves, the choice of who will operate the hotel is of paramount importance and can have a major impact on the returns enjoyed.

Unfortunately, choosing the 'best' operator for a hotel or resort is not straightforward. There is no single credible list that ranks potential operators from 'best to worst'. The vast majority of international operators are skilled when it comes to running hotels, and so it comes down to choosing who is most suitable for the specific property.

Many factors will come into play when determining who should operate the property including:

- the management fees and other costs involved in the agreement;
- capital expenditure requirements for converting or adapting the property for each respective operator;
- the level of experience for the type of property;
- the level of experience in the geographic region;
- the requirement for a new hotel in that region by each brand;
- the relative presence of the competing brands in the main source markets for the hotel;
- which brand will suit the hotel best, and help it to trade most effectively;
- which brand is likely to generate the highest EBITDA for the owner.

Sometimes the decision will not be easy. Is it better to employ someone who operates four similar resorts in the area because they have greater local knowledge? Or would that mean they would now have too many 'beds to fill', so the hotel might underperform? Would it be better to go with a resort operator with an excellent track record but no experience in that location? They could potentially bring their entire marketing machine behind the new location, enhancing trading. However, they might have additional regional office costs that may need to be met solely by your hotel.

It is vital that all options are explored fully if the most informed choice is to be made. Once the choice is made, it is important to remember that the right to terminate (if the agreement does not work) is an important option.

Note

1 A 'competitive set' will usually be made up of four or more similar quality hotels located close by the subject hotel. Ideally these hotels will have similar profiles to the subject hotel and similar facilities

3 Buying guide

What is a good buy?

A simple definition of a 'good buy' is a property that meets the needs of the investor, is reasonably priced and is affordable in the longer term. These three criteria are fundamental, although the relative importance of each one of these criteria may vary for different buyers.

Affordability

It is generally considered to be very important (by the purchaser themselves, their advisors and their funders – although less so by the vendor) that the purchaser does not pay too much money for their desired property. Any overpayment could well eat into capital that could otherwise be better utilised to improve (or maintain) the property. It could also mean that interest charges on the purchase loan consume a higher proportion of the earnings from the property.

From that perspective, it is vital to work to a business plan that is realistic to calculate how the business will work. This should not just be an exercise in raising finance from a bank, although it will be difficult to borrow money on a hotel without showing a sensible business plan. The exercise of writing the business plan should enable the purchaser to go through the whole thought process underlying the transaction.

The completed business plan should detail a number of key items including:

- why it is being bought;
- why the price is reasonable;
- where the current trade is coming from;
- where it is intended that future trade will come from;
- the proposed pricing structure;
- sales and marketing plans;
- staffing and resources plans;
- the strengths and weaknesses of the current property operation;
- the strengths and weaknesses of the purchaser as an operator;
- operational risks involved with the project;

- detail on the customer base;
- detail on the current and future competition, including SWOT analysis;
- changes to the trading environment;
- proposed repairs and improvement programmes.

Lenders will be keen to see a cashflow analysis for the property, which must detail, at the very least, the proposed expenditure (purchase price, and transactions costs, followed by future capital expenditure for any improvements or refurbishments) and a projected profit and loss account for the property down to EBITDA levels. The plan should then take into account the cost of interest on the loan and repayments to demonstrate that the debt will be affordable.

It is also sensible to undertake a sensitivity analysis to show what the impact of problems with trading could have on the loan, which will help the bank be confident that the business will be able to support the proposed level of borrowing.

All of these items should be looked into carefully, to ensure that the price being paid is sensible, affordable and will enable the buyer to achieve what is wanted from the purchase.

Meeting the buyer's requirements

There are many reasons why people invest in hotels, and the key reasons are as follows. One or more of these reasons can apply to hotel investors looking at a specific property.

To generate an income

Hotels generate income either from the rent that has been agreed with the tenant or from the operating profits of the business. The level of this potential income is one of the key reasons people choose one hotel over another, or indeed over another type of investment altogether.

To benefit from capital growth

Hotel investors are usually looking for capital appreciation. If they have invested in the right property in the right location, they would expect to see a reasonable level of capital growth, in addition to the income received, over the holding period.

To ensure capital security

Some owners purchase hotels because certain types of property are seen as relatively risk-free investments, in effect a safe place to invest their money, as it is unlikely to decrease in value.

To provide a new trading location

This requirement is relatively unique to hotels. Hotel operators are sometimes required to have hotels in different locations to meet the needs of their customers. As such, investment decisions are sometimes made to ensure they have hotels in key locations for the benefit of the overall business, rather than basing decisions on individual returns on a property-by-property basis.

The remaining criteria for a 'good buy' is 'reasonably priced'. This is often construed to mean cheap, or good value. However, the real meaning is that the risk/reward relationship is favourable to the buyer, meaning that the rewards outweigh the risk.

London estate perspective

Hugh Seaborn CVO, Cadogan Estate

At Cadogan we, like other London estates, have been recognising the positive impact hotels can make on an area: how they are an important constituent in raising awareness of a destination, defining the personality of an area, as well as providing an appealing offer to visitors and residents.

We have long been known for taking great care over selecting retailers who offer something special to the area, and are uncommon either because they are a domestic independent or an international brand that is new to the London market. Similarly, with restaurants we seek high-quality operators who take pride in what they do and thus contribute locally through their offer and reputation.

London estates have in the past tended to let hotels either on very long leases, perhaps ninety-nine years – which they would sell for a premium – or, in some cases, on shorter leases of, say, twenty years, which might be subject to a market rent. These approaches secured capital or income for the estate and understandably left the hotel to get on with their business with the ability to assign the lease if they wished to in the future.

By adopting operating agreements, an estate which has a long-term interest in the success and vibrancy of its area is able to carefully select the best partner as an operator to manage their hotels and maintain the quality thereafter. By this means, Cadogan has been able to apply the same rigorous estate management principles to ensuring that hotels make an impact on the place they are in.

This is perhaps a distinctly new chapter in the steady evolution of London estates from being ground rent collectors, as they were in the early twentieth century, to being involved and engaged with the business being undertaken in their properties. The increased operational risk has the potential to generate increased cashflow, subject to careful management. There is no doubt that it is necessary to have specialist expertise that understands the detailed dynamics of a hotel and restaurant business.

Figure 3.1 Trophy hotels are deemed secure investments

To understand the dynamics of this equation, it is important to first identify the potential risks and then ensure that the yield adopted adequately reflects the risk associated with the purchase.

Risk vs reward

Types of risk

There are many types of risk associated with hotel and resort investments, and these all need to be considered carefully:

- operational/business risk
- property risk
- location risk
- structural risk
- country risk
- political risk
- conditions risk
- competition risk
- economic risk
- supply risk
- currency risk
- demand risk
- the risk of a shift in investor perception/demand.

Figure 3.2 Safari lodges tend to be seen as riskier investments

The level and type of risk associated with the property will influence the appropriate yield.

Yield choice

The two key elements that affect the value of a hotel or resort are: how much income the owner will receive (EBITDA or net rent), and what multiplier (yield) is applied to that income stream. Terminology used in the hotel industry in respect of capitalisation rates (yields), discount rates and multipliers is similar to the rest of the property market. The appropriate multiplier is calculated by the following:

$$\frac{1}{\text{yield}}$$

So, for example, if the hotelier is prepared to pay 8 per cent then the multiplier they will adopt is 12.5 times.

$$\frac{1}{8\%} = 12.5$$

In this section we outline some of the things that affect yield choice. It has been subdivided into vacant possession and investment properties for ease of reference, although many factors apply to all types of properties.

Determining the correct capitalisation rate and discount rate with which to value a property is fundamental in producing an accurate valuation and, therefore, determining whether something is 'reasonably priced'. It is essential that every reference possible be made to market evidence, although there are many problems with such evidence in the context of the individual (and indeed often secretive) nature of hotel transactions.

The market generally discusses the yield upon which a property was purchased. This is calculated by taking the income (whether rent or stabilised EBITDA) and dividing it by the purchase price (and in the case of investments, deducting purchaser's costs).

Example: Hotel Vaughan

The Hotel Vaughan produces $1,900,000 EBITDA (stabilised trading) and was purchased for $24,000,000. The yield paid for the property was 7.9 per cent.

$$\frac{\$1,900,000}{\$24,000,000} = 7.9\%$$

When discussing the appropriate yield to adopt, we are trying to look into the mind of the potential purchaser for a property. The only way to determine what the 'reasonably efficient operator' is likely to pay is to consider who the bidders are likely to be for a particular property and to look into the factors that could affect their bid price.

Vacant possession properties

It is important to remember that the appropriate yield is the one that the successful purchaser would adopt when buying that specific property. It is likely that the market will consider many factors when deciding what multiple of earnings to pay.

The following are just a selection of factors that could have an impact on the yield applied when assessing the value of a hotel. Everything that could have an impact on the profitability of the hotel and/or future growth in the capital value is likely to have an impact on the appropriate yield.

Geographic location

The location of a property in terms of its surroundings is usually fundamental to its value. The yield for a hotel on the Champs-Élysées in Paris is likely to be sharper than the yield used for a similar hotel in Freedom Square in Tbilisi because of the greater demand from investors for the more established and, arguably, more secure market. In the same way, it is also likely that the yield will be sharper in a city centre rather than on a secondary roadside location.

Type of property

Whether the property is a deluxe five-star hotel or a bed and breakfast style of operation will have an impact on the yield that will be applied. However, it should be noted that it is not a straightforward descending scale with the sharpest yields being applied to the 'higher graded' establishments, as the capitalisation rate will depend upon the market demand for that category of property.

Quality and condition of property

The quality of the property may impact on the yield, as a hotel that is in better condition and requires less defensive capital expenditure will usually command a sharper yield than a rundown equivalent property.

Suitability of location for the property

If the property is located in an area to which it is not suited, then the yield is likely to be softer than would otherwise be the case if it were well located. For example, a conference hotel in a remote location may not attract as sharp a yield as one located in a highly visible location in the centre of a number of transport connections.

Market expectations and perceived desirability of the property in the current market

The underlying market sentiments towards a property will impact on the yield that property can command. For example, if the 'market' feels that mid-range hotels in Dubrovnik are under pressure from newer boutique hotels and more budget operations, then purchasers will not be prepared to pay as much for such a unit, unless it has evident development potential.

Level of current and proposed competition from investors for the property

The strength of demand from potential investors for a particular property will have a fundamental impact on the market yield. If demand is high for a particular type of property in a particular location, then a higher price will have to be paid to beat off the challenge of other purchasers. On the other hand, if there are lots of similar properties available at any one time, the yield may soften.

Level of trading demand

If the hotel market is particularly buoyant, then the number of new entrants into the market trying to gain a foothold will increase competition for the property, leading to a sharpening of the yield.

Historic trading profile of the property

The historic trading profile of the property will have an impact upon the yield adopted. In many cases, the more stable the trading history (and therefore the less risk associated with this element of the valuation), the sharper the yield profile adopted. However, where there is growth potential in the trading (for example, the hotel has been underperforming and the purchaser believes that they can improve the trading), then sometimes a sharper yield will be applied to reflect this growth potential.

Future growth in value

If an underlying growth in value of the property is likely (for example, if the country has just been accepted into the European Union and demand is likely to grow substantially), then a sharpening of the yield is likely to reflect the future residual value.

Business mix and the risk attached to various income streams

Arguably, if a substantial proportion of revenue comes from a less secure income stream (for example, F&B or short-term sublettings, rather than from rooms), then the overall capitalisation rate that it used will be slightly softer to reflect the additional risk to the income.

Over-reliance on one source of income

If one source of business generates too great a proportion of the turnover of a business, then the business starts to be reliant upon that other business, and the risk profile of the operation changes, potentially having an impact on the yield. For example, if a hotel were to take 60 per cent of all bookings from one corporate company, it would be at risk if anything were to jeopardise those bookings (for example, financial difficulties in the source business).

Changes in the economic environment

If an area's economic profile is changing, then it could impact on the appropriate capitalisation rate.

Changes in the local infrastructure

Any changes that could affect future trading are likely to be partially reflected in the selection of the capitalisation rate.

Alternative uses

Any alternative uses that could improve the value of the property in the future are likely to be reflected in the selection of the yield.

The possibility of 'treasure hunting' and releasing additional value

It is likely that if there are opportunities to improve the trading of the property through redevelopment, extensions or repositioning the property in the market place, then the purchaser will offer a sharper yield to secure the property.

Visibility and accessibility of the property

When visibility is important to the operation of the hotel, a higher profile outlook will often command sharper yields. Alternatively, if a property relies upon tourists for its custom and there is only one flight per week to the destination, the capitalisation rate may be lower than an equivalent property served by twenty flights per week.

Suitability for current use

If the property is not well suited to the use for which it is being purchased, then it is likely that a softening of the yield will reflect this.

Flexibility

The more operational flexibility the property has can be a commercial advantage for the operator and as such, this could result in a sharper yield being paid.

'Brand-ability'

If a property is run independently but is brand-able then it could mean that a higher proportion of people will be interested in buying the property (potential franchisees, for example), which could lead to more competition for the property and therefore a sharper yield.

Funding

The availability of funding and the provision of debt can have an impact on yields. Where high levels of debt are freely available, yields are likely to be sharper than when debt is limited and more equity is required to fund the purchase.

Yield selection for investments

Determining the correct yield with which to value a property investment is fundamental in assessing the value. It is important that every reference possible be made to market evidence.

However, in the hotel world there is no set formulaic standard for leases or management contracts that can be compared with 'institutional leases' that are commonplace in retail or commercial investments. As such, each investment needs to be assessed on an individual basis, drawing from the most similar types of other investment structures available.

Most of the criteria that is relevant for determining the yield for vacant possession properties is also relevant to investment properties (location, quality, economic environment, etc.).

In addition, there will be other factors to consider including the following.

Is it a lease or a management contract?

The nature of the investment vehicle will impact on the likely buyers for the property, and upon the appropriate yield.

Certainty of income

The more certain the income is, the sharper the yield is likely to be. As such, management contracts in unproven markets tend to attract softer yields than ground rents let to a government body in traditionally strong markets with strong underlying property values.

Who is the tenant (or manager, in the case of a management contract)?

The tenant will have a major impact on the yield that will be applied to an investment. If the same property is let to an international hotel operator with good financial standing, then more people will be prepared to buy the investment because of the perceived lower-risk profile, thereby pushing up its desirability and therefore the price (hence sharpening the yield). If the property was let to a local operator with no experience and no money, the yield would be softer to reflect the 'tenant risk'.

The financial standing of a company is often reviewed through various rating agencies like Dun & Bradstreet or Standard & Poor's. They provide risk assessments, which the investment market uses when assessing how desirable a tenant is likely to be.

Other factors like the quality of the operator, their suitability for the operation of the particular unit, their track record, their stated goals and the market perception of the operator will also have an impact on the market yield for such an investment.

Is there a guarantor for the rent and are they reliable?

Cases when the financial standing of a tenant is not as strong as an investor would like can have an impact on the yield. Either the price of the asset will need to be reduced (yield softens) to reflect the additional risk that the tenant will not pay the rent, or a guarantee can be provided.

In many countries, this can be a sum of money placed on deposit in a bank, with the investor being able to take this money in the event of default. In other cases, the provision of a guarantee from an entity with a greater financial standing can be provided.

As such, if a new hotel is being built, it may be let to a specifically created company (newco) but it may have to be guaranteed by 'the parent company' to provide the financial backing to increase its value.

Is the rent affordable?

Whether the rent can be paid by the tenant is a key consideration of what yield should be applied to a property. If the rent is too high, there is a much higher likelihood of default and as such, the price paid should reflect this additional risk (and the yield should soften).

Conversely, if the rent is substantially below the market level for a property of this type, the yield will sharpen to reflect the additional certainty of receiving the rent.

Proximity to vacant-possession value

A key comfort for most investors in a hotel is the underlying vacant-possession value of a hotel. If the tenant defaults on the rent, what price could be achieved if the property needs to be placed on the market with vacant possession? If the vacant-possession value is substantially higher or lower than the investment price, it may well have an effect on the appropriate yield to apply to that property, and therefore its value.

Impact of specific lease terms

Almost every lease clause will have an impact on the appropriate yield for an investment, as every term will impact upon the desirability and therefore the value of the investment. There are too many potential clauses contained within commercial hotel leases around the world to discuss all of these, but below we comment on a few key clauses.

Rent

There are many different types of leases that calculate the rent in various different ways, for example:

- fixed rent with fixed review periods to market rent
- fixed rent with fixed uplifts
- fixed rent with annual uplifts based on RPI
- turnover-related rents
- EBITDA-based rents.

The differing levels of certainty of rental income are usually reflected in the adopted yields.

UNUSUAL REVIEW PERIODS

Where existing leases provide for rent reviews at intervals different from the intervals currently being agreed for new lettings in the open market (some markets are usually reviewed annually, others commonly receive five-yearly reviews), it can sometimes have an effect on the yield applied to the investment.

LIMITATIONS ON USE

Where a user-clause is qualified to the effect that the use cannot be changed 'without consent, such consent not to be unreasonably withheld', there is unlikely to be a large impact on the yield adopted for the valuation.

Where the prohibition on change of use is absolute, there can be an impact on value. A restriction to a generic use such as 'hotel' will not normally have any valuation consequences. Absolute prohibitions that narrow the range of potential demand that are enforceable can adversely affect rental value (for example, if the clause specifies a use as a 'designer hotel').

RESTRICTIONS ON ALIENATION

An ability for tenants to be able to assign or sublet their lease is important, and a complete prohibition against alienation can have a severe effect on value. A hotel with a prohibition on subletting part of the property may also result in a discount to the value if such a restriction limits the flexibility of the tenant to let out surplus space to complementary uses.

ALTERATIONS

A prohibition on alterations, or the requirement for landlord's consent (the granting of which can be unilaterally withheld) may well impact on the desirability of the investment and hence the yield.

Impact of specific management contract terms

Almost every clause contained within the HMA will have an impact on the appropriate yield for an investment, as every term will impact upon the desirability and therefore the value of the investment. There are too many potential clauses contained within management contracts around the world to discuss all of these, but below we comment on a few key clauses.

THE 'TERM'

If the term is too long or too short, it could impact on the yield applied to the investment.

THE LEVEL AND DETAIL OF THE MANAGEMENT FEES

The cost of the management of the hotel by the operator will impact upon the cashflow generated by the owner, and as such will influence the value of the investment. It is usual to see a base fee (which is a percentage of turnover) and an incentive fee (usually a percentage of AGOP).

If the level of fees is not standard, it may have an impact on the yield applied. In addition, if fees are subordinated to debt (or other criteria), this may reduce risk and thereby sharpen the yield applied to the income.

PERFORMANCE TEST CRITERIA

Many management contracts contain no performance test criteria at all, in which case the enhanced operational risk is usually reflected in the yield applied. If the performance test is inadequate, it will also be reflected in the yield applied.

OWNER'S RIGHT TO SELL THE PROPERTY

Any clauses that impinge on an owner's right to dispose of their interest will have an adverse impact on the yield selected.

OTHER CLAUSES

Where the clauses contained within the HMA are non-standard, this may impact on yield selection. This can include the owner's right to approve budgets, appointments, control bank accounts, inadequate FF&E reserve provisions or restrictions on competition.

Other methodologies for assessing the appropriate yield: building up capitalisation rates without any direct market evidence

There are a number of factors that combine to determine the appropriate capitalisation rates that are used in the market; when no comparable evidence is available, these should be looked at to help provide an opinion of value.

These factors can include the following:

- cost of debt
- cost and availability of equity
- location risk
- property condition risk
- property category risk
- operational risk
- economic risk
- country risk.

This is a substantially more theoretical approach than using directly comparable evidence, and runs the risk of not necessarily reflecting the approach that would be taken by a potential purchaser. However, if done carefully, and if all the differing risk elements are accurately reflected, it can provide a reasonable starting point when no direct evidence is available.

However, where market evidence is available, this should always be referred to ahead of calculating the capitalisation rate from each underlying risk.

Another slightly more suitable method for building up hotel investment yields without direct market evidence may be to look at other investment classes available and to adjust the returns available elsewhere to reflect the nature of the property.

Alternatively, it may be appropriate to review similar transactions in similar geographic locations and then make relevant adjustments based on the differences in the area.

Operational due diligence

A potential purchaser or valuer needs to review the property to see how 'operationally relevant' the property is. A hotel, resort or leisure property needs to meet the needs of its potential customer base if it is to continue generating operating profits.

As such, it is essential that the due diligence undertaken reviews the operational suitability of the property for its given market. If a property is no longer suited to its historical market, repositioning the asset into a different market segment might be an efficient way to bring it back to operational relevance, without the need for extensive capital expenditure.

If the market requirement has changed, or is likely to change, then the achievable trading profits may be impacted, in turn impacting on the value of the property.

It is important to review each part of the business that makes up the overall property. For a resort, this might include an analysis of the bedrooms, the various bars, the restaurants, meeting space, spa and leisure facilities. If the product is only sold to resort customers, then it is important that the overall offering is relevant to that market. However, where non-resident guests are also attracted, then each segment must be relevant to its individual market place and bear favourable comparison with direct competitors.

A hotel needs to meet the specific requirements of the market it is trying to appeal to, allowing it to compete with current, and proposed future, supply if it is to stay relevant and successful.

From an owner's perspective, the hotel usually needs the ability to deliver bottom-line returns, while from an operator's perspective, it may also need to meet brand standards, to help build brand recognition for the operator and ultimately enhance the grand value.

Estimating the appropriate market yield where there is no market evidence

In certain markets, there is no comparable transaction evidence on which to base capitalisation/discount rates, although there would still be a market for a hotel if it were sold, making the 'profits method' of valuation the appropriate method.

There are a great many ways to create an 'all risks yield' from scratch, each of which will have its own difficulties. The key thing to consider is that, for the yield to be accurate, first it needs to reflect the approach the market would take, and second, each input needs to be accurate.

One method is to apply the following formula:

$$Y = B + L + P + T + O + E + C + Eq - G$$

where:
Y = all risks yield
B = base rate of lending
L = location risk factor
P = property condition risk factor
T = property category risk factor
O = operational risk factor
E = economic risk factor
C = country risk factor
Eq = availability of equity risk factor
G = growth

Another approach would be to adopt the following formula:

$$Y = P + i + r + d - g$$

where:
Y = yield
P = liquidity preference
i = inflation
r = risk
d = depreciation
g = growth

Other approaches apply different factors based on the availability and cost of debt, as well as hurdle rates for the necessary equity portion.

Whatever the approach, to be accurate it is essential to assess the likely market for such a property and then apply the relevant risk factors based on the market's perception.

What makes a great hotel bedroom?

There are two different ways to look at what makes a good or a great hotel bedroom: the customer's perspective and the owner's perspective. That said, in many ways they should both be aiming in similar directions. After all, the owner wants customers to be happy and to return, and indeed recommend their hotel to their friends, family and to be positive on TripAdvisor and other similar social media platforms.

The overriding principle is that the customer wants to have a value-for-money room, whether value for money is $50 or $10,000 for a night's stay. A clean, comfortable, safe, quiet room with a good bathroom and a good TV is the minimum standard traditionally demanded by guests, but times are changing. The first priority now for many travellers, both business and leisure, is connectivity. Free Wi-Fi is now essential, and indeed free 'fast' Wi-Fi is becoming standard throughout the industry.

In addition, one of the biggest complaints now of older hotels is that there are not sufficient, easily accessible power points for guests to recharge their phones, laptops, tablets and other devices. Ideally, a room should have a bank of at least four easily accessible plugs, preferably with at least one near the nightstand, for overnight charging of their mobile phone.

Other facilities that help improve the quality of the room from the customer's perspective include: having a good hairdryer with a good mirror and decent lighting, sufficient hanging space with good clothes hangers (not the 'thief-proof' ones that make customers feel that the hotel does not trust them), good bathroom accessories (including dressing gowns) and good lighting in the main chamber of the bedroom.

Ideally, the room will have windows that open, with a good view, and a mini-bar that allows space to keep drinks or even children's meals chilled.

There are two complaints that keep recurring within the industry, but are often not heeded by hotel developers. First, the ability to keep charging devices when the customer is not in the room is seen as a great help by most guests, so potentially designers need to apply the power-saving key slots just to the lighting, TV and air-conditioning units, leaving the power sockets off the loop. Second, non-magnetic door keys are definitely in favour, as anyone who has got to their thirtieth-floor bedroom only to discover their phone or tablet has managed to deactivate their key card will attest to.

However, things are slightly different from an owner's perspective. They obviously want the room to meet the customer's needs, so that booking will be strong. They also want the best product (for their chosen market segment) at the most reasonable price, so that investment returns are highest. In addition, they need the room to be designed efficiently, so it can be serviced as quickly as possible, and any repairs and replacements can be made at minimal cost.

The design of a bedroom can save up to 40 per cent on servicing times if the layout is carefully thought through at the design stage. Ensuring the right TV channels are in the room can lead to customer satisfaction (if the clientele is primarily from the Middle East, adding TV channels suitable for that market is a sensible approach), while removing the sports channels and only showing them in the bar can sometimes lead to additional F&B spend in the bar.

However, whatever is done to the bedroom, it is essential to provide the key basics correctly. A dirty room, an uncomfortable bed, a poor bathroom or a lack of hot water are all unacceptable to the vast majority of hotel customers and will lead to complaints, unhappy customers and a poor reputation. It does not matter how good the service, bar, restaurant or spa are – without these basics being addressed, the hotel will fail as a business enterprise in the long term.

4 Development guide

David Harper and
Mark Martinovic, CEO, Hotel Spec

Development process – the road map for good development

In normal circumstances, the process of developing a hotel follows a logical sequence, as shown in Figure 4.1.

Secure the site

The first step of the development process is to secure the site on which the development will be built. Each site will have different characteristics that lend themselves to different types of development, so without knowing the specifics of the site, you cannot start to work on the develop process.

Also, if all the work developing a concept is completed without having the site secured, the promoter could quite easily be wasting their time and money. Owning the site is not essential if a secure 'purchase option' can be negotiated.

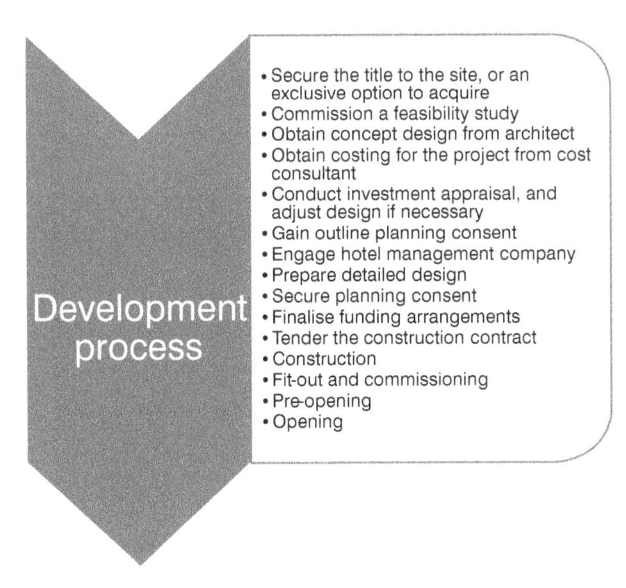

Figure 4.1 Development process

Commission a feasibility study

The second step is to undertake a feasibility study to ensure that, first and foremost, a hotel will make sense on the site. In addition, the study should help to determine the optimum type and size of hotel to be developed on the site.

A specific location will dictate the optimum type of hotel, and sometimes developing only 200 bedrooms on a site will not make the most of the site's potential, while at other times 200 bedrooms will be too many bedrooms for a location to sustain.

Obtain a concept design from an architect

The purpose of the concept design is merely to allow a cost consultant to price the development in general terms. The design is likely to change at a later stage of the development process, so too much design work at this point is likely to be uneconomical.

Development costings

At this stage, only rough parameters for the construction should be obtained. These will obviously change as the final design is undertaken and the construction contracts are tendered, but at this stage of the process outline costs are required.

Investment appraisal

The costs of the development should be compared with the value of the proposed completed hotel. If there is an insufficient profit margin in the development, the scheme should be reviewed to see if sufficient profit can be generated by the development to match the risk profile. If the profit margin is not 'adequate', it might make sense to alter the project, put the development on hold or potentially cancel the project altogether.

Secure outline planning consent

Advice should be sought as to whether the proposed development plan would be acceptable in outline terms to the responsible government authority. In principle, would a hotel of the proposed quality and quantum fit in with the zoning of the land, and would a detailed planning application potentially be favourably received by the planning authority?

Engage a hotel management company

This is the stage to decide who will be the best operator for the property. This is typically done by engaging a specialist agent to hold a 'beauty parade' among the most suitable operators. The chosen operator should be selected, and then the best terms for the management contract should be negotiated.

Detailed design

Once the management company has been determined, the final design can be undertaken. Without an operator on board, there is a strong likelihood that the design will need to be altered once they are on board. Design criteria specific to the operator will need to be factored into the final design.

Secure planning consent

Once a detailed design is in place, detailed planning consent can be sought. This can be a lengthy process, and it is always useful to engage a planning expert to ensure the process is made as easy and as smooth as possible.

Finalise funding

Once the developer has the site secured, a feasibility study in place, a management company lined up, projected costs and a detailed design, funding can be sought. At this stage, it is possible to provide enough information to lending institutions to ensure that suitable financing terms can be agreed, subject to a detailed construction cost tender document.

Tendering the construction contract

The next step on the development process is to pick the best construction company for the project, preferably through an open-tender process. The analysis of tender offers will take into account the experience of the construction company, timescales, guarantees, penalty clauses, quality and number of subcontractors, as well as the final tender price.

Construction

The construction can then begin. Construction needs to comply with local legislation, regulations from the finance institution, and the specifications of the management company, if all parties are to be happy with the process.

Neglecting to meet these obligations can have very severe penalties. Not meeting local building regulations can lead to the property needing to be demolished. Not abiding by the terms of the loan can lead to funding being withdrawn and early repayment being required, potentially causing the developer to lose their entire investment.

Failing to meet the specifications of the management company can lead to the management contract being terminated, potentially leading to the loan being withdrawn. With so much at stake, the whole process should be managed by an experienced project manager to ensure the construction phase goes smoothly.

Fit-out and commissioning

The final finishes of the hotel are of vital importance for its trading potential, so care should be taken when procuring the goods. Typically, the operator and the interior design team will want to have a sizeable input into this stage of the process.

Pre-opening

As completion gets nearer, the operator will move into the pre-opening phase of the development. They will start recruiting staff, training staff, agreeing suppliers, as well as starting to market the property. Business travel agents and local companies will be contacted to talk about the hotel and perhaps even invited to come and have a look at what will soon be available. Sometimes the pre-opening phase will include a partial (or 'soft') opening, where only part of the hotel's facilities are opened, allowing the staff to learn the layout and style of the hotel without the pressure of a full letting inventory.

Opening

The last stage is for the hotel to open and become fully operational, earning large amounts of income for the owner. Over their first year of operation, it is typical that small construction glitches come to light, and it is common for a final payment to be withheld until these 'snagging items' are rectified.

How to minimise problems in the development process

Planning ahead

A detailed development plan, overseen by an experienced project manager, can help ensure that each part of the development is carefully coordinated to minimise unnecessary delays and problems.

An experienced team

The importance of having the right team cannot be over-emphasised. Some problems that occur on a project may be unique, but the vast majority of problems will have been encountered many times before and an experienced team can deal with most problems quite comfortably, having seen and rectified them before.

Finance available

Having the development finance available from the outset of the project ensures that the project can be planned to run smoothly, without delays occurring due to

Figure 4.2 Site in Honduras

the need to secure further funding. The Radisson Blu in Lagos, Nigeria was finally opened almost eight years after it was meant to be opened; the single most important reason for this delay was a lack of development funding during the process.

How to maximise returns

It is relatively obvious to say that building the right hotel (for the site), for as low a price as possible (without compromising the required quality) and ensuring the best management for the property are the first steps to maximising the developer's returns.

However, one factor that is not always taken into account is the quality of the construction company used. Some well-known construction companies can unlock development funding that would otherwise be unavailable. Sometimes the funder will offer preferential rates as their perceived risk is reduced by having a reliable and well-known construction company on board. The same can be true when employing certain project managers.

One of the key determinants of a developer's return which is often overlooked is the construction period. It goes without saying that the sooner the development is completed, the quicker an income can be produced, but the impact of even minor delays in opening times on overall returns is quite marked.

As such, it is really important to price-in any delays to the construction period when looking at the completed tender documents from various construction companies. The overall construction cost might be 15 per cent higher for one firm but the speed of delivery might mean that the overall developer's returns are higher, even with the higher construction cost.

Figure 4.3 Site in Angola

Feasibility studies

What is a feasibility study?

As the name suggests, a feasibility study is an analysis of the viability of an idea or concept. The feasibility study focuses on answering the essential question, 'should we proceed with the proposed project?'

Promoters with a business idea should always conduct a feasibility study to determine the viability of their idea before proceeding with the development. Learning early on that a business idea will not work saves time, money and heartache later.

A feasible hotel development is one where the business will generate adequate cashflow and profits, withstand the risks it will encounter, remain viable in the long term and meet the goals of the founders. It will also be worth more upon completion than it costs to develop.

A feasibility study is not a business plan. The separate roles of the feasibility study and the business plan are frequently misunderstood. The feasibility study provides an investigating function. It addresses the question of, 'Is this a viable business venture?' The business plan provides a planning function and it outlines the actions needed to take the proposal from 'idea' to 'reality'.

Why should a feasibility study be carried out?

Promoters and developers will often consider saving time and money by not undertaking an independent feasibility study; this is almost always a mistake, and sometimes a very expensive one.

Project promoters may find themselves under pressure to skip the 'feasibility analysis' step and go directly to building a business. Reasons given for not doing a feasibility analysis include:

- We know it's feasible; an existing business is already doing it!
- Why do another feasibility study when one was done just a few years ago?
- Feasibility studies are just a way for consultants to make money.
- The market analysis has already been done by the business that is going to sell us the equipment.
- Why not just hire a general manager who can do the study?
- Feasibility studies are a waste of time. We need to buy the building, tie up the site and bid on the equipment.

These reasons should not dissuade the promoter of a project from conducting a meaningful and accurate feasibility study. Once decisions have been made about proceeding with a proposed business, they are often very difficult to change. The promoter will need to live with these decisions for a long time.

The feasibility study is a critical step in the business assessment process. Completion of the feasibility study and valuation appraisal is an essential early step in the development process, and is required for the following main purposes:

- to provide information to the promoter of a project that it is viable;
- to determine the most appropriate facility provision in light of current and projected future market conditions;
- to provide a basis for a financing structure, including the proportions of debt and equity which can be supported by the project's cashflows;
- to act as an introductory document for potential financial partners;
- to submit to potential hotel management companies, and to act as a basis for contract negotiation.

In addition, other benefits of undertaking an independent study include:

- giving focus to the project and outlines alternatives;
- narrowing business alternatives;
- identifying new opportunities through the investigative process;
- identifying possible reasons not to proceed;
- enhancing the probability of success by identifying and addressing mitigating factors early on that could affect the project;
- providing reliable information for decision making;
- providing documentation demonstrating that the business venture was thoroughly investigated;

- helping secure funding from lending institutions and other monetary sources;
- helping to attract equity investment.

What is involved in a hotel or resort feasibility study?

The feasibility study for a hotel or resort is quite detailed. The study will usually be performed in three main phases:

- Phase 1 – site appraisal
- Phase 2 – market and general appraisal
- Phase 3 – financial appraisal.

Phase 1 – site appraisal

At the beginning of the study, the consultant will visit the hotel/site/property to assess the accessibility and position relative to: other hotels and resorts in the area, the airport, commercial and leisure facilities that could generate demand, and other places of relevance in the area.

They will consider the suitability of the property for its intended use, and identify its strengths and any weaknesses in that regard. This will then be used as part of their analysis to determine the likely impact on future trading, given a number of development options.

Phase 2 – market and general appraisal

During and after the site visit, the consultant will:

- obtain current and historic market and operational data on the hotel industry that is relevant to the study, including levels of demand, seasonality and market segmentation, arrival statistics and other market data;
- determine the existing and projected characteristics of demand markets, including estimates of growth which are considered to be of relevance to the project;
- determine the present supply of hotel accommodation serving those markets, and review any plans for new provision which would provide competition in the future;
- obtain other available information relating to the economy, communications and general development in the area, and assess the economics of, and general climate for, the operation of hotels there.

During the course of the study, the consultant will conduct interviews with: a selection of operators of existing hotels in the area; representatives of the conference and travel trade; airport and airline officials; representatives of local industry and commerce; and other parties who can provide information on the markets for a new hotel in the area, including planning and development information and pertinent economic and other background information.

They will also speak with relevant governmental officials or their advisors, including planning officers, as well as the various parties of the design team for the specific project.

Subsequent to the site visit, research into potential demand-generating sources for new hotels in the area, and the project's likely market acceptance and penetration, will be carried out. This market research will be used to help formulate the recommendations regarding the opportunity for developing/redeveloping the hotel or resort, its market positioning and the facilities it should offer.

Phase 3 – financial appraisal

Based on the research and the conclusions drawn regarding location, competitive situation, market positioning and facility provision, the consultant will project room occupancy and average room rate for the hotel's accommodation facilities, along with levels of utilisation of, and revenue from, conference, F&B, leisure and other facilities over the first five or ten years of operation.

A statement of estimated profit and loss for the period up to and including the stabilised operating year of the project, based upon an estimated opening date, will be prepared, generated from all the research that was undertaken. These estimates will usually be prepared in accordance with accepted uniform accounting practices, and the assumptions upon which the estimates are based should be detailed within the report.

What is the process of instructing a feasibility study?

Pre-feasibility study

A pre-feasibility study is usually conducted by the promoter/developer first, to help identify a few of the relevant scenarios, for example, size parameters and the quality of the proposed development. A consultant can help the developer with the pre-feasibility study but ideally the promoter/developer will be involved. This will then be used to frame the parameters of the feasibility study.

Feasibility study

The promoter/developer should then look for the best consultant with relevant market knowledge and experience to carry out the feasibility study.

Review the results and conclusions of the study

The conclusions of the feasibility study should outline, in depth, the various scenarios examined, and the implications, strengths and weaknesses of each. The project leaders need to scrutinise the feasibility study and challenge its underlying assumptions to ensure that they are the most sensible assumptions to make.

I don't need a feasibility study – this is a dead cert!

Trevor Ward, managing director – W Hospitality Group

Okay, good luck – after all, you will need lots and lots of it! You're walking into a minefield without a map, a map which is readily available (for a fee!), and which is now, and throughout the development period and beyond, useful for everyone involved in the project.

It's not easy to foretell the future, which is what you need to do in a feasibility study. And in these days of massive external shocks, such as the Gulf War, 9/11, the global financial crisis, Ebola and the most recent oil price collapse, it gets even more difficult. But ironically, such impacts are the easiest to factor-in to a future projection, simply by modelling a decrease in revenues every five or so years in the 'pessimistic' scenario.

Any feasibility study is basically a supply and demand analysis. Who wants to stay in your hotel? And what product – type of hotel, size of rooms, price point, ancillary facilities, etc. – do they want?

Did you hear the one about the developer who, in the absence of any advice, decided to build The Ritz in a secondary city in Nigeria where the most expensive hotel room was US$200 per night, and he needed to charge US$600 for his rooms to make it work? 'They'll come', he said, 'it will be so wonderful they will want to stay there'. They came, they saw and they stayed next door, at the hotel built by someone who, with the benefit of good advice, knew the city would never attract enough people who would pay US$600 but could attract overnight guests who would pay US$250 for a product better than the existing hotels.

It is not the purpose of the feasibility study, or the role of the consultant, to decide whether or not to proceed with the business idea. It is the role of the project leaders to make this decision, using information from the feasibility study and input from consultants. The decision to go ahead or not to go ahead is one of the most critical in business development. It is the point of no return. Once the promoter has definitely decided to pursue a business scenario, there is usually no turning back without potentially high abortive costs.

Operator selection – the value of the brand

Do brands add value?

Within the hotel industry, there is much talk about the benefits of brands and how they help a hotel trade better, and thereby add value to the real-estate component of the property. Most hotels are owned by one party (the investor), but then the hotel may be managed by a hotel brand, or potentially be franchised-out, so it can trade as a branded hotel. As such, the 'brands' own very few hotels and rely upon the value of their brand to stay in business.

If instructing a brand to manage the hotel did not generate more 'value' for the owner, then they would be unlikely to go down that route. In the same way, if adopting a franchise (that obviously has cost implications) did not add value, then the owner would be unlikely to follow that through.

All brands can show examples of where their brand has added significantly to the performance of an unbranded hotel, suggesting that they do indeed add value to the value of the underlying real estate.

However, as many hotel agents can tell you, having the ability to sell the property with vacant possession can have a positive impact on sales prices, suggesting that a brand does not always add value. To add real value to the real-estate component, either the EBITDA (after all management fees have been deducted) needs to be higher, or the yield applied to the property needs to be sharper with the management team in place than without them. As such, it is worth exploring the areas where a brand can add value.

Marketing

The major hotel brands spend a lot of money promoting their brands so that customers know what they offer and are happy to stay at one of their hotels, knowing what they can expect. One 'oft-quoted' customer said, 'When I wake up in the morning I may not know which city I am in, but I can always tell that I am in a Holiday Inn'.

The standardisation of hotels has become so clear-cut that in most brands you can rely upon the service levels, the quality of the hotel fittings and even the majority of the items on the menu in the restaurant.

This homogenisation has an impact on customers' expectations and can create (when it is done well) brand loyalty. As such, when a hotel is associated with a brand, it will ensure it benefits from better brand awareness than it would do if it were an independent hotel.

Management

A brand on a hotel can have a positive impact on the management of the hotel. If the hotel is managed by a brand, then that brand will ensure that a fully trained manager is in place, and that the hotel runs in an efficient manner.

If the property is franchised, the brand will only allow a franchise to be granted if the proposed management is good enough (in their opininion) to manage one of their hotels. As such, in certain areas, some brands won't allow many third-party owners/operators to take franchises as they are uncertain of the quality of the management team, and they are wary of the damage a poorly managed hotel can do to their brand.

Changing online profile marketing

Will Hawkley, director and UK Sector Head for Leisure and Hospitality – KMPG LLP

A few months ago, I stayed in a hotel in Oxford on business. I booked through an OTA and had a satisfactory stay. This was the first time I had ever stayed in Oxford and I doubt I will need to stay there again in the foreseeable future.

However, since then I have received approximately two emails per week from the OTA, offering me special deals on hotels in Oxford. As I sit on a plane about to fly to New York, not once has the OTA sent me an email about hotels in that city.

It is obvious to see that they are using crude historical data to push me irrelevant offers that sit in my spam filter and get deleted. This is annoying, to say the least, but also a complete waste of money on their behalf.

They would provide a much better service and customer experience if they could use predictive analytics to gauge my future travel patterns, rather than just pushing me offers that are not relevant.

The technology exists to enable this and to map customer journeys from beginning to end, across multi-channels and multiple devices, but companies need to invest in data science quickly or they will get left behind.

Companies need to use not just in-house data (if they are able to collate and analyse it, which many companies can't) but also the multitude of external data sources that are available on the Internet and in the social media world we live in.

Some people may think that this is too invasive and that customers will push back but I would rather receive relevant information and offers than useless ones, and the GenYs and Millennials are completely comfortable with this use of data as they are very happy to post it on numerous social media platforms, especially if it allows them to travel more efficiently and cheaply.

In conclusion, it is clear to me that companies need to invest in data science, not only to get ahead of their competitors, but also to deliver the best possible customer experience from end to end.

GDS

The global distribution system (GDS) is vitally important for a hotel as it allows customers to find and book rooms online, in real time. All brands imply that their GDS is better than any other, but there is no way to be certain which is better; indeed, it is likely that the benefits of each system will vary depending on their specific geographic locations. However, what is clear is that having access to GDS is better than not having access to it. The brands all provide a hotel with direct access to a GDS system, helping gain bookings for the hotels listed on their system.

Loyalty schemes

Loyalty schemes have had a large impact on customer spending patterns, and the most successful brands have managed to really benefit from these schemes, to the advantage of all the hotels within the brand.

The typical scheme awards 'points' for each booking, and when a customer has collected enough points they can redeem them for some type of benefit, for example, a free stay or a room upgrade.

The impact of such schemes has led to OTAs like Hotels.com, Booking.com and even Airbnb to offer loyalty schemes. An independent hotel has limited opportunity to match such schemes, so not being part of a larger brand can have an adverse impact on bookings, because of the loyalty scheme alone.

Service levels

When a customer books into a Four Seasons hotel, they know exactly what to expect. They know they will get a warm welcome, will be escorted to their luxuriously appointed room, shown the facilities (if that is what they want) and will be provided with every luxury they could desire. An Embassy Suites hotel will provide them with a more basic level of service – they will still receive a friendly welcome, they will find their apartment is a standard size, that there is a bar and a manager's 'welcome hour' where drinks and snacks are provided free to guests. The Holiday Inn Express experience will be more basic still: a simple, basically fitted-out room, with a bar adjoining the reception and, potentially, one of two meeting rooms.

As customers, we all expect to be valued and will not forgive rude or unresponsive staff, but if we are staying in a limited-service hotel, we do not expect a wonderful spa, swimming pool and casino.

On the other hand, if a customer is staying at a hotel that charges a premium rate, then they would be disappointed to find only a simple hotel with no additional facilities. This is one of the difficulties that fashionable boutique hotels face: that after the initial interest in the property dies down, if the quality of the offering does not match up to the price charged, trade can decline.

The 'brands' ensure that the facilities and service levels match brand standards, so customers know what to expect.

FF&E reserves

Brands insist upon an FF&E reserve being put in place, to ensure the property has enough money to keep it in good condition for its customer base. Such prudence ensures that the hotel does not get unduly devalued, as the quality of hotel is kept up, allowing the hotel to trade at a suitable level, rather than having to cater to lower-rated business as the quality deteriorates through lack of investment.

Independent hotels may sometimes neglect to invest in the FF&E as regularly as is required, especially in times when the trading environment is challenging. Such a 'saving' will have an adverse impact on trade in the medium and long term.

Borrowing money

In many new developments, and sometimes when a hotel is being purchased, lending institutions feel that some of the operating risk is dissipated if a well-known operator is involved, and so they offer better lending terms than would otherwise be offered for an unaffiliated hotel. Indeed, sometimes they will not consider lending at all without brand involvement.

Sharper capitalisation rates

There is evidence to show that in less-established markets, real-estate buyers are attracted to branded hotels, and indeed in certain markets the very presence of a brand attracts more potential purchasers for a hotel, thereby enhancing the multiplication factor applied to the property, increasing its value.

Areas of insecurity or un-established markets

When a hotel is located in an area which is perceived to be unsafe, having a brand adds comfort to customers, especially those new to the location. Such first-time visitors tend to opt for brands they are familiar with for their first visit. If the city or country is considered dangerous, customers believe that the brand will ensure hot running water to the showers, will make sure that guests are not kidnapped from their rooms and will provide at least a club sandwich on the room service menu. Of course brands cannot ensure these things in all situations, but they will always use their best efforts to do so, which provides enough security to most nervous customers.

In more 'insecure areas', the premium attached to brands becomes extremely high, even where the danger level is not actually raised from more established cities.

Where do brands add little value?

However, the cost of becoming involved with a brand, whether just looking at management fees or franchise fees, means that in a number of situations the brands do not automatically enhance trading over and above these additional costs, and so do not automatically 'add value' to the real-estate component of the business.

Established markets

The internet has been an invaluable tool for hotel customers. When booking a hotel in London or New York, there is, thanks to websites like TripAdvisor, no need for the customer to rely on a brand to determine the likely quality of their chosen hotel. Repeat visitors to a location are more likely to look at independent hotels when detailed information is easily available, which has led to a lowering of the benefits of a brand in such locations.

Indeed, in such locations an oversupply of hotels of one brand can actually lead to lowering ADRs as the brand attempts to offload unsold rooms.

Boutique hotels

Boutique hotels are a classic example of where a brand adds very little to the value of an asset, although many brands are now attempting to move into this market.

Experience hotels

In recent years, there has been a dramatic increase in the demand for what have become known as 'experience hotels'. These are hotels where the primary purpose of the booking is 'the experience', rather than to stay in a hotel room in a certain area, which is the traditional motivation behind most hotel bookings. Someone might book a luxury safari lodge, or a stay at a Disney castle hotel, or even book a night in a lighthouse or a yurt. Such bookings are usually made purely because of the property, and the location of the property is of secondary importance to the customer.

In such circumstances, it is arguable that having a brand on such a property would not enhance the customer's likelihood of booking the stay, and might even have a negative effect.

How to assess which brands will add most value

To some degree, assessing which brand will add most value to the property is best left to an expert. Unfortunately, choosing the 'best' operator for a hotel or resort is not straightforward. There is no single credible list that ranks potential operators from 'best to worst'. The vast majority of international operators are skilled when it comes to running hotels, and so it comes down to choosing who is most suitable for the specific property.

Many factors will come into play when determining who should operate the property, including:

- the management fees and other costs involved in the agreement;
- capital expenditure requirements for converting or adapting the property of each respective operator;
- the level of experience for the type of property;
- the knowledge and experience in the geographic region;
- the requirement for a new hotel in that region of each brand;

- the relative presence of competing brands in the main source markets for the hotel;
- which brand will suit the hotel best, and help it trade most effectively; and
- which brand is likely to generate the highest EBITDA for the owner.

Sometimes the decision will not be easy. Is it better to employ someone who operates four similar resorts in the area because they have greater local knowledge? Or would that mean they now have too many 'beds to fill', so the hotel might underperform? Would it be better to use a resort operator with an excellent track record but no experience in the location? They could potentially bring their entire marketing machine behind the new location, enhancing trading. However, they might have additional regional office costs that would need to be met solely by your hotel.

It is vital that all options are explored fully if the most informed choice is to be made. Once the choice is made, it is important to remember that the right to terminate (if the agreement does not work) is an important option.

Designing the hotel

Hospitality projects carry a higher financial risk than most other types of developments. As such, creating designs that are not unduly extravagant is usually important in ensuring that a project is viable. Architects and designers should understand that the design needs to be appealing, while reinforcing the brand identity and providing safety and security. But it should also maximise the revenue and profit potential of the hotel. The substance of a hotel (whether it is fit for purpose) is generally more important than the style in the first instance. It is a waste of money to design something that is pleasing to the eye but difficult to operate and maintain, making it more costly to run.

To design and build a successful hotel, the architect and the owner need to understand how a hotel functions. From a small fifty-bedroom luxury hotel to a 500-bedroom budget hotel, the basic organisational structure, roles and responsibilities of the staff remain the same.

The following basic functions have to take place. Someone must:

- be able to receive the guests and check them into their room
- prepare and serve food and beverages to the guests
- clean and replenish the rooms
- handle the accounting functions of the hotel
- maintain the hotel
- manage all these staff and functions.

An overall sense and knowledge of how a hotel functions is a critical tool that an owner should employ early on, and this is what should also guide the designer.

There are three distinct areas referred to in a hotel – guestrooms, front of house (FOH) and back of house (BOH). FOH refers to all the areas that can be accessed by guests and BOH is exclusively used by the staff of the hotel.

Having stayed in many hotels around the world does not qualify an owner or a designer as a hotel expert. They need to understand the relationship between the above areas, as well as the spatial planning, ergonomics and interaction between these areas from an operational perspective. Poor planning of the BOH areas, for example, will invariably lead to higher operating costs and poor guest satisfaction.

Heating and cooling space – it's not all air conditioning

Alistair Brooks, managing director – Commercial Real Estate Services Limited

Air conditioning or air handling has become a standard for most hotels above budget-level around the world, almost irrespective of the prevailing climate. However, there is often confusion between air conditioning and air handling.

Air conditioning

Simplistically defined as 'a system or process for controlling the temperature, humidity, and sometimes the purity of the air within a property'. 'Conditioning' of the humidity and purity is the key differentiator. The most common systems will centrally heat/cool and condition air and then duct this air around the building. Air conditioning is likely to be a market staple, rather than a premium in locations with high/low humidity or contaminants/dust or in trophy hotels.

Comfort cooling/heating

Pure heating/cooling without 'conditioning'. This can be heated centrally and air-ducted to rooms. Most modern systems heat/cool a medium either centrally or close to the room. This is transported to in-room equipment which transfers the heat/cooling to the room's air. The latter can be further broken down into:

- 2-pipe (i.e. 1 in/1 out) – this is less expensive to operate and install in a central system but (depending on scale) more expensive if set up as one system per room. The downside of a 2-pipe centralised system is that the whole system will be in either heating or cooling mode.
- 4-pipe (i.e. 2 in/2 out) – this has separate heating and cooling circuits feeding a single exchanger unit in the room. This provides greater control, enabling different rooms or zones to be heated while others are cooled.

The professional team

The professional team for a new build hotel usually includes the following primary consultants:

- project manager
- architect
- interior designer
- mechanical and electrical (M&E) engineers
- structural and civil engineers
- planning consultant
- quantity surveyor
- landscape architect
- fire and life-safety consultant
- security consultant
- kitchen, laundry and BOH consultant.

Additional secondary consultants may be required depending on the design, scope and hotel brand of the project. The following are additional consultants that might be required for typical hotel developments:

- IT/ELV (Wi-Fi, IT, telephone systems, TV systems, CCTV, etc.)
- lighting
- audiovisual
- signage
- building façade engineering
- procurement.

Note that under a design and build, or turnkey, contract, the above team of professionals is not required. The promoter may decide to hire an architect to design a concept for a hotel and then hand over that concept and all specifications as part of the 'employer's requirements'. The contractor will develop the concept design into full construction drawings.

Project manager

The project manager is required to lead the professional team in the design stages of the project and ensure that coordination between all the various parts of the professional team and construction team are correctly conveyed on the construction drawings.

Different forms of contract refer to this position with different terms but the roles and responsibilities are essentially the same. Under the FIDIC (Fédération Internationale des Ingénieurs-Conseils) and NEC3 (new engineering contract)[1] forms of contract, the project manager is referred to as the 'independent engineer', while under GCC (general conditions of contract), it is referred to as the project manager.

The role of the project manager is to ensure that the design standards, criteria and requirements of the employer and the hotel operator are all taken into account and provided for in the design works. Depending on the form of contract, the project manager should ensure that the design is measured in a detailed bill of quantities, under a traditional form of contract, and collated with the contract documentation, ready for issue in the tender process.

As part of the tender process, the project manager should prepare a list of previously pre-qualified contractors who shall be invited to tender for the project. Once the tender documents have been received, the project manager will present all tenders to the promoter and will assist them in adjudicating all the bids and then preparing the contract documentation for final agreement and signature.

Once the contractor has been appointed, the project manager's role then involves overseeing the process of the physical construction and monitoring the 'conditions of contract' to ensure that all parties adhere to their obligations.

On completion of the project, the project manager will oversee the testing and commissioning of all plant and equipment, oversee the compilation of snag lists, receive and review final accounts, and certify completion of the works.

In order to successfully achieve the above objectives, it is crucial that the appointed project manager has sufficient skills, qualifications and experience in dealing with construction projects, but also sufficient knowledge and experience of hotel projects.

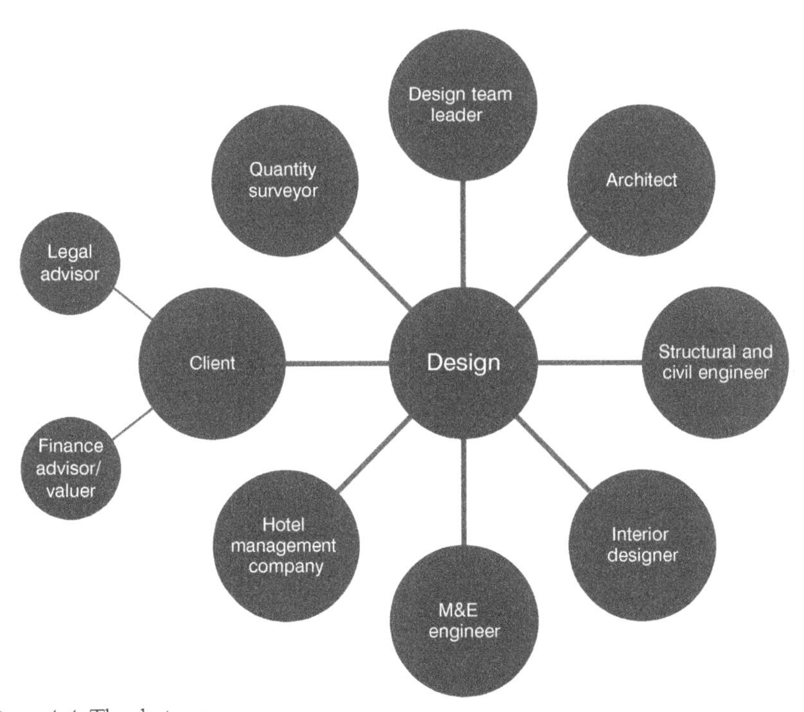

Figure 4.4 The design team

Architect

The architect is usually the first consultant appointed. In appointing an architect for a hotel project, it is vital that the promoter employs an architect with experience in designing hotels.

The architect should establish the rights, constraints, consents and approvals for the particular site that the hotel is to be constructed on, in order to establish the project brief. The architect must also take into account the promoter's requirements for the hotel, and where a hotel operator has been appointed, obtain a list of all of the brand requirements in terms of the architectural planning and guidelines.

The key criteria that they will abide to when designing a hotel are as follows:

- *Operations* – meeting the needs of the guest efficiently and at the required service level. Additionally, ensuring that the property can be cleaned and serviced efficiently and cost effectively.
- *Visual appeal* – ensuring that the design appeals to the target market, bearing in mind the quality and the price point of the hotel.
- *Repairs and maintenance* – ensuring that the design of the property allows the property to be easily kept up to standard, with the choice of FF&E readily available to be replaced if needed.
- *Initial capital outlay* – the cost of the development must be in line with the quality of the finished hotel or resort.

The architect will work closely with all the other members of the design team to produce detailed plans and layouts of the hotel or resort, deciding on the form and materials of the finished product.

Interior designer

The interior designer has a very important function in the design of the hotel, and can sometimes find themselves in conflict with the architect, if they are correctly fulfilling their brief.

They are responsible for internal area relationships and need to put function ahead of aesthetics, at least until the spatial design works well. They will ensure all F&B outlets can be serviced adequately by the kitchens, that the room design is easily cleaned and that all public areas are welcoming to the guests. They will ensure there are adequate BOH areas to service the hotel. They will also advise on lighting throughout the property and will help decide on the décor and general image of the hotel.

Mechanical and electrical engineers (services)

M&E services often become a bottleneck in the design stage of a hotel project. The simple reason for this is the sheer number of services required by a hotel.

Consider the amount of air conditioning, extraction and fresh air required by the building; the amount of water services (hot, cold and sewage) to all the bathrooms, kitchens, laundry, etc.; and all the electrical services throughout the property. Add to this the ELV service which provides the phone lines, TV signals, internet facilities, CCTV monitoring, etc. This all creates a huge network of pipes, cables and air ducts throughout the property, which need to provide services to every area and be able to be maintained throughout the life of the building.

All these services need to be designed seamlessly into the building and incorporated and coordinated with the architectural, structural and interior designs, to ensure both pleasing aesthetic design while still being easily accessed for maintenance.

Structural and civil engineers

The structural and civil engineers undertake the initial site inspection so that any site-specific factors (soil type, gradient, geology, etc.) can be taken into account by the architect in the design process. They are responsible for ensuring the structural integrity of the building design and will advise the architect on the minimum structural requirements for each element of the design.

Quantity surveyor

The quantity surveyor is responsible for costing the design and, as such, has a position of vital importance. It is their responsibility to ensure the development costs are kept to a reasonable level so that the development is viable and can go ahead. They need to negotiate the prices for materials and even insist on design changes to ensure the development budget does not exceed affordable levels.

Landscape architect

The landscape architect is often the forgotten member of the design team, and when costs need to be saved, this is often an area where cuts are made. However, this can be a mistake, as the skills a good landscape architect brings can lead to an excellent first impression for hotel guests, enhancing the all-important 'sense of arrival' that often leads to higher rates being generated as well as enhanced repeat business.

Hotel operator

In all cases where an international hotel brand has been signed up, the operator will sign a technical services agreement with the owner, whereby the operator assigns a technical expert to advise the owner and guide his design team in providing the necessary standards for that particular hotel brand.

In instances where the final HMA has not been signed, most international operators will still sign the technical services agreement in order to assist the owner in the design programme.

The standards include specifications and requirements related to construction and design, which should be respected in order to meet the standards in terms of quality, style and comfort for that particular brand.

The main objectives of implementing these standards can be summarised as follows:

- to ensure the safety and security of guests and employees;
- to deliver consistent quality, comfort and service to hotel guests;
- to increase operational profits through: the efficient use of space; reduced consumption of utilities; the selection of reliable, durable and well-designed products which are of an approved quality; meeting the functional requirements; and creating an optimum balance of style against price;
- to establish specification guidance and acquaint the owners and developers with their construction and design obligations in building an internationally recognised branded hotel.

These standards do not aim to cover every single detail of the design and construction of a hotel property, but set out the minimum requirements to be followed while designing and/or constructing a hotel, and they are subject to the regulations of local government building control and/or planning authority statutes.

The operator will normally assist the owner throughout the development and pre-opening stages of the hotel and should be relied upon for their experience and knowledge.

Note

1 This is the latest updated version of a different form of contract. It is similar to FIDIC but has its own form, more popularly used in the UK.

5 Construction guide

Mark Martinovic, CEO, Hotel Spec

Identifying, understanding and managing risks in hotel construction projects

In order to succeed in the development of a hotel project, one needs to understand the various risks associated with hotel developments.

Procurement is a term which describes the activities undertaken by a developer who is seeking to bring about the construction or refurbishment of a building. It is a mechanism which provides a solution to the question, 'How do I get my project built?'

On most projects, clients (usually through their in-house advisors or third-party professional consultants) will start the procurement process by devising a project strategy. The strategy entails weighing up the benefits, risks and budget constraints of a project to determine what the most appropriate procurement method is, and what contractual arrangements will be required.

With every project, the client's concerns focus on time, cost and quality (the performance or value 'triangle') in relation to both the design and construction of the building.

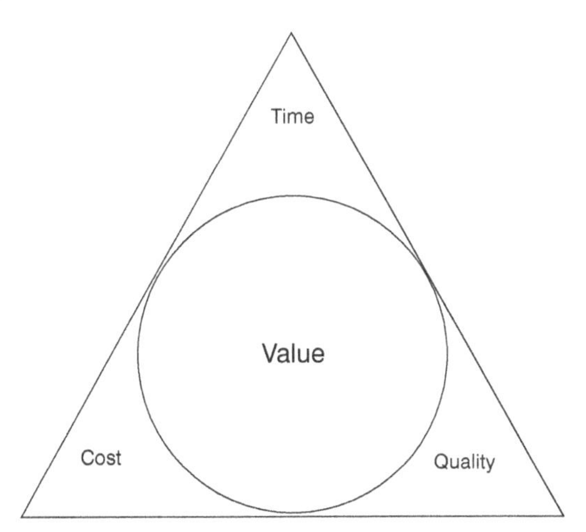

Figure 5.1 The value triangle

This value triangle is not fixed and in fact is quite pliable, in that any increase or decrease in any one of the three performance factors will directly impact on the others.

The difficulty, and what sets construction industry procurement far apart from anything else, is the complexity of projects. Influences such as ground conditions, topography, logistics, weather, available technologies, finance, labour availability and services, to name only a few, all affect the ability of a project to be completed on time, on budget and to a high quality.

The client's resources, organisational structure and preferred contractual arrangements will all need to be taken into account in choosing the right procurement method for their project.

Understanding risk is essential, as although each procurement method follows a well-established set of rules and procedures, there are risks associated with choosing any particular route.

Successful procurement relies on all parties involved in the project complying with their respective obligations, and identifying and dealing with risk appropriately from the outset.

Procurement, being a series of risks, has different methods that transfer varying levels of risk between the client (referred to as the 'employer' under the Fédération Internationale des Ingénieurs-Conseils – FIDIC – forms of contract) and contractor. Plainly put, the various types of contracts deal with how the risks are shared between the employer and the contractor. Typically, the more risk that is transferred to the contractor, the higher the price for the contract will be.

The conditions of contract are the terms that collectively describe the rights and obligations of contracting parties (i.e. the employer and the contractor) and the agreed procedures for the administration of that contract.

Typically, these conditions address the following:

- the parties' main responsibilities, e.g. the employer provides the site and the right of access thereto, while the contractor provides the works in accordance with the requirements established in the contract;
- the timing of the works, including the start date, time for completion, period for defects, liability, etc.;
- testing and remedying of defects;
- payment, e.g. the manner in which the works are to be assessed and certified, time for payment, and interest on overdue amounts;
- variations and claims, e.g. the manner in which variations to the contract are to be evaluated and paid for, and how the costs which result from employer liabilities are assessed and paid for;
- title (ownership) to objects, materials within the site, etc.;
- risks and insurances, e.g. what are the employer's and contractor's risks, and what insurances each party will take out;
- termination, e.g. the reasons for termination, the procedures for termination and the payment to be made upon termination;

The following are examples of standard forms of contract used in Africa when engaging main contractors for construction works contracts:

- FIDIC (French initials for 'International Federation of Consulting Engineers') (1999) (Rainbow suite of contracts – Red, Yellow, Silver, Gold and Pink Books).
- General Conditions of Contract for Construction Works (GCC).
- New Engineering Contract (NEC3) (Engineering and Construction Contract and Engineering and Construction Short Contract).

- the resolution of disputes, e.g. will disputes be determined by adjudication, mediation, arbitration, litigation (court of law) or a combination thereof?

Conditions of contract can be standardised so that the same conditions of contract can be used on different projects, in which case they are referred to as standard forms of contract.

Forms of contract

Traditional contracting

Traditional contracting (hereafter referred to as 'traditional'), which can be described as the separation of design and construction using a lump-sum or re-measurable contract, has its weaknesses, as all methods of procurement do. However, the construction industry has used the traditional process for so long that it has become the best understood. This almost universal understanding of traditional is its greatest strength – the consultants are responsible for design and the contractor for execution, so responsibility for coordination of subcontract packages lies firmly with the contractor, while responsibility for design coordination and ensuring sufficient design and specifications information is provided to the contractor lies with the client/employer.

The 'design and build' and the 'engineering, procurement, construction/turnkey' (EPC/T) models both seek to address the 'division of responsibilities' issues faced when using the traditional contracting route. Both offer a single point of contact for a developer in respect of the design and the construction of the project. Additional specialist involvement, such as design consultants and/or proprietary processes, are all engaged by the contractor as subcontractors, which again preserves the single point of responsibility.

By having one entity appointed from the outset to perform the design and the construction, these two models, outlined here, allow the design and construction to proceed almost in parallel, which can often deliver the finished project quicker than traditional contracting is able to.

Design and build (D&B)[1]

In the D&B form of contract, the contractor designs and provides, in accordance with the employer's requirements, 'the works'. The works may include any combination of civil, mechanical, electrical and construction works as specified by the contract. This is a useful form of contracting insofar as the contractor undertakes all of the design works as well as the physical construction. The major benefit in employing this form of procurement is the reduced input and design coordination required from the client/employer.

It is important to stipulate in detail the 'employer's requirements' when entering this form of contract. This includes all the finer details, for example, incorporating the hotel operator or brand requirements, specification of finishes and other such requirements. It is not sufficient to tell a contractor that you wish to build 'a Hilton hotel, like the one at Frankfurt airport', as the contractor is not aware of all the detailed specifications of the plant and machinery, the size and requirements of the back of house, fire and life safety standards or the minimum performance standards of all plant and machinery that are required by Hilton.

The employer's requirements is a document included in the contract that specifies the purpose, scope and/or design and other technical criteria for the works. The contractor therefore is only required to provide a building that functions according to these specifications and design criteria.

It is highly recommended that a client/employer employs an independent interior design consultant under this form of contract in hotel projects, as typically contractors are not interior designers and it goes without saying that once a fixed-price contract has been concluded, the contractor will try to optimise cost-saving measures in order to reduce his costs and increase his profit. Therefore, for example, if you require natural stone tiles to the lobby and 100 per cent back-up diesel generator power, then these have to be clearly stipulated in the employer's requirements at the outset.

As with the EPC/T form of contract below, the optimal situation for a D&B contract would be to employ a concept architect and interior designer to prepare conceptual drawings and schedules of specifications to illustrate what the desired finished product should look like and contain. These conceptual architectural and interior designs would then form the basis of the employer's requirements.

The hotel brand would then add all their minimum specification requirements and you would have a very good starting point for the detailed employer's requirements. By going this route, the contractor would be provided with a lot more detail in order for him to provide a comprehensive cost and time plan, as well as understand the standard and quality of finishes required. Thereafter, the contractor would provide all the engineering design and construction details for the construction, while the client/employer's project manager or engineer administers the contract, monitors the construction works and certifies all payments.

Turnkey (EPC/T)

Engineering, procurement, construction/turnkey has become a preferred form of contract with lending institutions, particularly with a fixed price (or lump sum) where the total cost of the contract is stipulated in the contract and understood from the outset. In this form of contract, the majority of risk is placed upon the contractor with the exception of *force majeure* and certain other unforeseen risks.

This form of contract may be suitable, and indeed is preferred by international lending institutions, as it provides first, a higher degree of certainty of final price and time, and second, the contractor takes total responsibility for the design and execution of the project with little involvement from the client/employer.

Employers using this form must realise that the employer's requirements which they prepare should describe the principle and basic design of the project. In hotel developments, this can be quite complex in terms of the facilities requirements of the hotel, but also of the minimum brand standards imposed by the hotel operator. If these requirements are not sufficiently covered in the employer's requirements, then the contractor may be entitled to claim for additional works or changes to the design to comply with the hotel operator's requirements.

This is almost always the outcome where the design has begun without the hotel operator being signed up. When the hotel operator has been appointed, certain minimum brand standards have to be followed. This could include bigger bedrooms, additional F&B facilities, increased safety requirements or a host of other changes to what the contractor was initially provided with in the employer's requirements.

In this form of contract, the contractor should be given the freedom to carry out the work in his chosen manner with little or no interference from the employer, provided the end result meets the performance criteria specified by the employer. The challenge here can be that the hotel operator (under a hotel management agreement and technical services agreement) retains the right to sign off at all stages of design; also, many hotel developers want to follow the progress of the development and be assured that the programme is being followed and they also want to monitor the quality of the works.

Under the usual arrangements for this type of contract, the contractor carries out all the engineering, procurement and construction: providing a fully equipped facility, ready for operation (at the 'turn of a key').

Table 5.1 is a simplified table outlining the roles and responsibilities under an EPC/T contract. This is not a complete list but it does address many of the major contractual issues. The way each of these issues is handled can be modified during contract negotiations to suit the situation and overall goals of the project.

EPC/T contracting tends to be more expensive to the client/employer, due to the shift of project risk away from the client/employer and to the EPC/T contractor. On average, a project will cost 10 per cent to 20 per cent more using EPC/T style of contracting than a project using the more traditional design, measure, tender style of contracting. This is due in large part to the project's risk being more evenly distributed between the client/employer and contractor.

There is a misconception among clients and contractors with regard to the definition and nature of the following forms of project: FIDIC EPC/T projects; the Silver Book; Plant and Design-build; and the Yellow Book (first edition, version 1999 for both),[2] which are leading to disputes and conflicts and therefore major

Table 5.1 Roles and responsibilities under EPC/T contracts

Task	*Role and responsibility*
Equipment supply contracts	Negotiated and signed solely between EPC/T contractor and supplier.
On-site construction contracts	Negotiated and signed between EPC/T contractor and supplier.
Supplier selection	Suppliers chosen solely by EPC/T contractor with no input from owner.
Scope of supply	EPC/T contract only as good as the original project specifications presented during bidding process. Changes to specifications/scope of supply after awarding of contract can be expensive, due to EPC/T contractor's sole contract with employer and employer's inability to 'shop around' for multiple quotations from independent contractors/suppliers.
Equipment supply warranties	Warranties negotiated by suppliers and EPC/T contractor and issued to EPC/T contractors directly. Warranty to owner from EPC/T contractor is negotiated separately between owner and EPC/T contractor and issued to owner by EPC/T contractor.
Process warranties	Warranties negotiated by suppliers and EPC/T contractor and issued to EPC/T contractor directly. Warranty to owner from EPC/T contractor is negotiated separately between owner and EPC/T contractor and issued to owner by EPC/T contractor (usually a performance bond).
Construction site safety (general liability insurance, workers compensation, accident, etc.)	Site safety solely the responsibility of the EPC/T contractor and sub-contractors; in accordance with contractual agreements.
Permitting (environmental, construction, etc.)	Permitting is the responsibility of the EPC/T contractor with the exception of permits that are required by law to be issued in the name of the owner of the project.
Project budget cost overruns	The cost risks for a project are borne by the EPC/T contractor. Any cost overruns, for equipment and/or services within the EPC/T contractor's scope of supply, are for their own account and cannot be passed onto owner unless 'change conditions' occur or contractual agreements to the contrary.
Project budget cost savings	The cost risks for a project are borne by the EPC/T contractor. Any cost savings, for equipment and/or services within the EPC/T contractor's scope of supply, are for their own account and are not passed onto owner unless contractual agreements to the contrary.
Project day-to-day expenses	The day-to-day expenses for the project, within the EPC/T contractor's scope of supply are borne by the EPC/T contractor.

delays and obstruction to contractors' activities. This will eventually lead to legal claims and variations between clients and contractors which are unfavourable to both parties. Therefore, both parties should seek advice and understand the true nature, rights, responsibilities and conditions set by each type of contract and their suitability to a hotel construction project.

It is essential to stress that the contractor will be fully responsible for the design and construction of the plant in both types of contracts. However, the employer can have very little involvement (and may not interfere in the contractor's activity) with an EPC/T contract where the contractor will submit documents and materials for information only. This is in contrast to the employer's solid control, review and approvals for all contractors' documents in D&B contracts.

There are two main differences between the EPC/T and D&B contracts in the general conditions of contract.

Liability and risk

In the EPC/T form of contract, the contractor will accept total liability and responsibility of the project to deliver a functional hotel. This will lead to increased costs of the project to cover the higher risk of unforeseen events that might arise during construction, where the employer normally issues only the employer's requirements along with major equipment specifications and a soil investigation report, and will leave the complete design responsibility to the contractor. D&B contracts, on the other hand, will divide the risk and liability (as stipulated within the conditions of contract) between the contractor and the employer. The level of shared responsibility will depend on the level of design details completed by the employer and issued in the tender document. The employer can issue any level of design details, ranging from simple employer's requirements, to a concept design or a detailed list and/or manual of minimum requirements and performance specifications from a hotel operator. Under general design obligations, the contractor is given a design-review period in order to scrutinise all employer's documents calculated from the commencement date, and to notify the employer of any error, fault or other defect found in the employer's documents. The employer can then decide whether to accept the contractor's notification or proceed as per the original design at his own risk. Following this period, the contractor will adopt the employer's design and take full responsibility for it.

Role of the engineer[3]

A further difference between the two types of contracts is the role of the engineer assigned by the employer to supervise and control the contractor's activities in a D&B contract. The engineer will represent the employer and will be authorised to exercise the rights assigned to him by the employer under his contract. The engineer will have the right to issue instructions on behalf of the employer, review and approve materials and the contractor's documents. A very important

The importance of getting clear instructions from the client – Four Seasons in Moscow

The Four Seasons Moscow opened in 2014 after the old building had been demolished in 2004 and replaced with a purpose-built design. The previous building was the Hotel Moskva, which was built between 1932 and 1938 and designed by Alexey Shchusev, with the intention of becoming one of Moscow's finest hotels. However, the design was notable for two very differently designed wings coming from the central core.

It is reputed that Shchusev presented the two different options to Stalin for approval, and the General Secretary of the Communist Party signed off both designs accidentally, not having noticed that they were different. The story goes that the architect was too scared to inform Stalin that he had failed to select a design and instead decided to build the hotel exactly as laid out, with two differing designs.

Figure 5.2a The original Moscow Four Seasons hotel in 1975

Figure 5.2b Moscow Four Seasons hotel in 2015

sub clause in the D&B contract is sub clause 3.5 [Determinations] where the engineer is given the power to agree or determine any matter that arises during the project between the employer and contractor by consulting each party. If agreement is not reached, the engineer has the right to make a fair determination in accordance with the contract, taking due regard of all relevant circumstances.

EPC/T contracts, on the other hand, will allow the employer to assign an employer's representative with limited authority to act on his behalf, and monitor the contractor's activities and ensure that it is done in accordance with the employer's requirements and specifications. The role of determination under EPC/T is given to the employer who will make a fair determination on any disputes, in accordance with the contract. This can often lead to further disputes, as the employer is not an independent party and is involved in the dispute. Therefore, the employer's determination could be deemed to be unfair and/or difficult for the contractor to accept.

It is vital for employers to pay very close attention to the details in the employer's requirements and specifications issued in an EPC/T contract as any missing specifications will entitle the contractor to install any item that he considers 'fit for purpose' without regard to the employer's best interest.

Employer's requirements

It is critically important that the employer's requirements sets out in sufficient detail what the end result of the project should entail and what is expected by the employer. It is not recommended that this document is submitted in the tender documents prior to the appointment of a hotel operator and a specific hotel brand. The simple reason for this is that each hotel brand has its own set of guidelines and brand standards, and many of the larger international hotel brands have extensive performance specifications for various plant and equipment, as well as detailed requirements for the finishes, facilities (both front and back of house) and fire and life safety (FLS), among others. Most international hotel management agreements have specific clauses requiring the owner/developer to comply with all of the brand standards for the specific hotel brand they have selected. This is a contractual requirement (in terms of a signed management contract), whereby the owner has to produce a fully equipped and furnished hotel to the standards required by the operator.

Where a hotel developer has decided to go ahead with his development prior to appointing a hotel operator, and then selects an operator and hotel brand during construction, there is a very high risk of having to make a number of changes to the design, specification of plant, equipment, furnishings or spatial planning. For example, an operator may require more than one service elevator for a hotel larger than 100 rooms and if only one was planned for then structurally and architecturally, this will have major implications to the cost. In addition, there will be time delays for the design work to be completed and then additional works to be effected onsite. The result of this would be the contractor being able to submit an 'extension of time' claim; there will be additional costs for design and then the purchase of the additional elevator, which in itself has a long order lead time and is an expensive item in any building. Along with the extension of time claim from the contractor, he will be permitted to add 'preliminaries and general' (P&G)[4] costs to each day of extension and will be entitled to reasonable profit on the addition to his scope of work. Even in a fixed-price, lump-sum contract, this late adjustment to the design and/or the addition of equipment or built-up area will result in additional cost and time to the employer.

Finances

Client/employer

The employer needs to ensure that they have secured funding for the entire project, as delays in payment to the contractor can have a severe impact on the project. A contract for construction will require an advance payment, normally 50 per cent of the total contract value.

Afterwards, there are monthly payments to be made. Some developers are under the impression that once 50 per cent of the contract sum has been advanced, that they do not need to pay again until the development is 50 per cent completed. This is not the case and the contractor will submit regular claims according to an agreed schedule of payments. These payments will have deducted from them the retention amount, as well as 50 per cent of each claim (the amount of the advance payment); however, they will still have amounts owing to the contractor which must also be paid.

Failure to pay the contractor on time will entitle the contractor to compound interest on the outstanding amount. The risks here are the increased interest charged by the contractor to the employer, adding to the project costs but also – where a smaller contractor has been employed – this may put financial pressure on the contractor and inhibit his ability to perform according to the contract. In addition to the interest charges, the contractor may suspend works due to non-payment, or they may have to delay ordering certain equipment that requires upfront payment from them, thereby delaying the contract to which they would be entitled to a claim for extension of time.

Contractor

During the tender stage, and in the tender documents, it is important to require all tendering contractors to provide a copy of their balance sheet to verify their financial ability to execute the project. The contractor should be required to submit an 'advance payment guarantee' from a reputable bank prior to payment of the advance payment. They should also be required to submit a 'performance bond' and agree to a suitable retention amount, which will be addressed later.

Under an EPC/T contract, the employer is not responsible for payment to any other contractor, supplier or subcontractor as that all becomes the responsibility of the contractor. It is important to know that the contractor is paying their subcontractors and suppliers on time so there is no risk of delay by others, as that will invariably affect the overall time programme of the project.

Insurances

There are essentially three phases of a greenfield project that need to be considered, together with the appropriate insurances and risk-mitigation measures.

Phase 1– pre-construction

- Professional indemnity insurance needs to be provided by all professional service providers, e.g. architects, engineers, quantity surveyors and project managers. Minimum limits of indemnity of US$5m per claim for engineers and US$2m per claim for all others are advised. Often a consultant will state that their professional indemnity insurance will be to a maximum of double their fees.
- Preferably the client/employer should take out a project professional indemnity policy that will provide cover as above for a minimum of six years for the entire professional team. Fees to professionals are then reduced proportionately for their contribution to the insurance. Banks and other financiers much prefer this arrangement and it provides the employer with the certainty that all insurance policies are current.
- Performance guarantees must be provided by the contractor prior to commencement of work, usually for 10 per cent of the contract price (banks or insurance companies provide this security).
- A project risk assessment and evaluation should be conducted on each project.
- Under an EPC/T contract, the employer will not appoint individual professional consultants as the contractor takes total responsibility for the design and execution of the project.

Phase 2 – construction phase

Insurances should preferably be controlled by the client/employer as described above, but often the contractor will provide his own insurance certificates as required by the project.

Main insurances should include:

1. contractors' all-risk up to total contract price plus 30 per cent escalation;
2. public or third-party liability for a minimum of US$5m per claim;
3. lateral support (if required for adjoining property risk);
4. project delay, delayed start-up, covering increased financing, working and building costs up to a minimum of US$10m (or as agreed by owners and funders);
5. contractors must provide proof of plant, motor and employee injury insurance.

Phase 3 – operational phase

Main insurances required are:

1. assets all-risk covering any loss or damage to the buildings, infrastructure, equipment and contents; insured up to full new replacement value plus 30 per cent escalation;
2. business interruption/loss of profits insured up to full gross profit/income less variable cost, or as agreed by owners, funders and hotel management company;
3. comprehensive public liability (specialist cover required for the hospitality sector); the minimum recommended value is US$10m.

It is suggested that the client undertakes a project risk assessment or employs a qualified service provider to ensure that all risks and exposures are addressed appropriately. Again, it is important to engage with the lenders on the project, and most of them will have their own minimum requirements for insurances.

Other risks

Delays caused by authorities or changes in legislation

Where certain changes have come about as a result of a change in import procedures or delays by authorities, and the contractor can prove that they had diligently followed the procedures laid down by the relevant legally constituted public authority in the country of the project, then the contractor will be entitled to an extension of time for completion.

In the event of any change in legislation that occurs after the contract has been signed and the result of that change in legislation is a change in cost, be it upwards or downwards, then the contract sum can be adjusted according to the increase or decrease, as the case may be. If the change in legislation causes a delay to the contractor's progress, then the contractor will be entitled to a claim for extension of time. Along with the extension of time, again, the contractor will be entitled to payment of P&Gs.

Client-supplied items

In many hotel projects with private individual owners, the owner elects to procure and supply certain items to the contractor. In the first instance, it should be clearly stated that the items to be supplied ('free-issue material') shall be delivered free to the contractor onsite. This means that the client/employer will arrange all necessary freight, insurance, clearing and forwarding costs, etc. so that the contractor receives the items onsite at no additional cost to himself. Alternatively, the parties could negotiate that the client/employer selects the supplier of a product and the contractor imports the specified product from that supplier at his cost, which will be itemised in the bill.

The biggest risk here is that some hotel developers believe that they can procure items cheaper from China. In most cases, the client is not a specialist in the particular product and since it is a client-supplied item, the contractor will not accept responsibility for the performance of those items and will not cover the warranty of such.

A further risk is the ability of the client/employer to ensure that all free-issue materials are delivered to site on time in accordance with the programme of works. If the delivery is late, partial or items are damaged – causing a delay to the programme – the contractor will be entitled to claim for extension of time. Where the items arrive too early, they will need to be securely stored in an environmentally appropriate warehouse giving due regard to rain, humidity, dust, etc., the cost of which should be considered.

Long-lead items

Unfortunately, sometimes plant and machinery items can be 'long-lead items', meaning that it takes a long time to get them to site. Before they can be ordered, they must also be designed, specified and agreed by the hotel operator. The ordering, supply, logistics and local clearing and delivery to site must be carefully coordinated in order to avoid any delays to the project's programme.

Other issues could be: problems in port, international exchange rate exposure, customs procedures, availability of transport vehicles, state of the roads and infrastructure, for example. In addition, there could be problems with the availability of local specialist installers and the availability of spare parts and service personnel.

Plant and machinery are not the only potential long-lead items. Much of the FF&E and M&E items required by international hotel brands are not all available in the local markets, so these may also have to be imported with the same risks and perils as above.

Quantities of FF&E and M&E for a 200-bedroom hotel will not necessarily be kept in stock and will need to be manufactured, so this additional time must all be factored into the programme.

Another classic risk with long-lead equipment or plant is the equipment warranties. Potentially, the lifts need to be ordered early to have them arrive onsite and be installed in the lift shaft before the building can be closed up. All reputable lift manufacturers offer a warranty on their equipment, but by the time the construction has completed and the hotel has opened, the warranty period may have expired, so now the lifts are no longer covered under the equipment manufacturer's warranty and any equipment failure now becomes the owner's problem.

Unknown conditions

Typically, under an EPC/T contract, the contractor takes full responsibility for having foreseen all difficulties relating to the project and by signing the contract, acknowledges having obtained all information as to the risks, contingencies and other circumstances which could influence or affect the works.

However, if fossils, relics, articles of cultural, national, geological or archaeological interest are discovered on the site and the employer is compelled to instruct the contractor to delay works, or provide for the excavation of such findings, then the contractor shall be entitled to an extension of time and payment of any additional costs in excavating such fossils or other articles.

Force majeure

Force majeure means an exceptional event or circumstance:

1 which is beyond a party's control;
2 which such party could not reasonably have provided against before entering into the contract;
3 which, having arisen, such party could not reasonably have avoided or overcome;
4 which is not substantially attributable to the other party.

Force majeure may include, but is not limited to, exceptional events or circumstances of the kind listed below, so long as conditions 1 to 4 above are satisfied:

1 war, hostilities (whether war is declared or not), invasion, act of foreign enemies;
2 rebellion, terrorism, revolution, insurrection, military or usurped power or civil war;
3 riot, commotion, disorder, strike or lockout by persons other than the contractor's personnel and other employees of the contractor and subcontractors;
4 munitions of war, explosive materials, ionising radiation or contamination by radioactivity, except as may be attributable to the contractor's use of such munitions, explosives, radiation or radioactivity;
5 natural catastrophes such as earthquake, hurricane, typhoon or volcanic activity.

If the contractor is prevented from performing any of its obligations under the contract as a result of *force majeure* and if it incurs any costs in relation to the *force majeure* event, then they will be entitled to an extension of time and payment of any such costs.

Disputes

A dispute can arise in the contract when both the contractor and the employer think that they are right, and they both think the other party is wrong. There are a variety of reasons for a dispute, but the most common reasons are regarding money or time for completion.

The contract specifies the manner in which disputes will be resolved. The different forms of contract deal with disputes differently and typically include provisions for the initial resolution of a dispute by:

1 *Mediation* – a third-party is appointed to facilitate conciliation between the parties and agreement on the outcome of the dispute.
2 *Adjudication* – a third-party is appointed to make a decision on the dispute, which is binding on the parties and is final unless it is reviewed by either arbitration or litigation.

Where a party to the contract is dissatisfied with the decision of an adjudicator or fails to settle the dispute through adjudication, the dispute is referred for final resolution to:

1 *Arbitration* – one or more arbitrators are appointed to resolve the dispute in accordance with procedures laid down in the arbitration procedures of a recognised arbitration body nominated in the contract.
2 *Litigation* – one or more advocates are appointed by both parties. The advocates present their sides of the argument in a court of law to persuade the judge which party has the better legal case.

Mediation and adjudication are intended to provide an inexpensive means for resolving disputes without legal representation. Only if these procedures fail to resolve the dispute is the matter referred to arbitration or litigation for final resolution. The procedure for the adjudication of disputes is intended to resolve disputes quickly. Most contracts have time periods for referring disputes for resolution by an adjudicator. Disputes that are notified outside of these time periods are automatically referred to arbitration or litigation.

Not all disputes end up being declared as such. Most disputes are settled on-site between the contractor's representative and the employer's representative. The parties come to an agreement to settle the dispute in an amicable way, and all parties compromise on their position until they reach a common ground. This is the most effective way of sorting out differences.

Variations

Invariably, all hotel development projects will have some form of variation or other prior to the completion of the project. As good as the design team may be, actual works onsite may reveal a conflict in the coordination between certain trades; the hotel operator may have updated certain standards; or the owner decides to introduce an additional bar, restaurant or spa, for example.

A hotel developer is entitled to initiate a variation at any time during the contract prior to the completion and taking over of the project.

Having received an instruction to vary, the contractor under these forms of contract is required to submit a proposal for any modifications to the programme, together with his proposal for the adjustment of the contract price (to include the cost of the modification or addition). One should note again that even in a fixed-price contract, any instruction to vary will have a consequential cost and time implication. This can include the possibility of a reduction in the scope of work. For example, if a tennis court is removed from the contract, for possible budget or operator requirement reasons, then likewise the employer should benefit from the cost and time saving it would have taken to construct the tennis court.

Completion

'Completion of the works' generally occurs when the buildings reach a state of readiness for occupation of the whole development, although some minor work may be outstanding. Completion of the contract occurs when all obligations have been discharged.

Completion of the works is usually followed by a defects liability period during which the contractor is obligated to make good any shortcomings in the materials and workmanship covered by the contract that are indicated by the employer or his representative (engineer or project manager).

The completion of the contract cannot take place before the expiry of the defects liability period.

Completion of the works triggers the release of performance bonds and the reduction in retention monies. The completion date for a contract is usually linked to the completion of the works. Failure to do so may result in liquidated damages (informally referred to as penalties). Consequently, it is in the contractor's interests to complete the works as soon as possible.

Each of the forms of contract has different administrative procedures in bringing the works and the contract to completion. It is very important that contractors and employers familiarise themselves with the procedures that apply to their specific projects.

A defects or 'snag' list is typically issued by the employer's representative around the time that completion of the works is certified. Such a list, depending on the form of contract that is used, will typically indicate what needs to be completed prior to certification of completion, or what needs to be corrected during the defects liability or correction period.

A final walk through the works should be arranged for all the parties involved in the contract just prior to the end of the defects liability or correction period to confirm that all defects have been attended to.

When work is planned and designed, many assumptions are made that may not be correct when it comes time to construct. As such, the works may be constructed differently from how they were originally designed. The employer must have these details correctly recorded for future maintenance and possible additions or alterations that may be needed later on.

The contractor is required to hand over operation and maintenance manuals which enable the hotel operators and their engineering team to operate the plant that is provided or to take care of and service such plant.

The contractor's work is not complete until they have provided all this information in the format that the client requires.

Defects liability

'Defects notification period' is a negotiated period of time after the employer has taken over the building during which the snag list must be attended to and all defects remedied by the contractor. A defects notification period should normally be not less than twelve months for a hotel project, as with an extensive building such as a hotel which has a large amount of MEP plant, a large built-up area, and various functions that need to operate 24/7, it is not always easy or possible to identify all defects within a shorter period. The defects notification period is negotiated in the contract prior to signing and can be anything agreed to between the parties; so this period could be longer or shorter than twelve months.

At the end of this period, the retention monies have to be returned to the contractor if he has remedied all the defects

Retentions

Retention monies are an agreed amount that will be held back from all payments until final completion. The employer must check on any conditions precedent from their lender, as often the bank will stipulate a minimum amount (expressed as a percentage) to be held in retention. This often becomes an area of contention with the contractor as the retention monies are held back until the end of the defects notification period.

Attention should be paid to the detail in the bills as there is always the risk of 'item loading'. Item loading is a situation where items in the bill are overstated in terms of their rate-per-unit, and will usually be the items in the bill that are used very early in the construction phase. For example, a contractor could overstate the cost-per-cubic metre for concrete, or cost-per-kilogramme of the steel which goes into the foundations. The result of this would be that the contractor makes excessive profit in the very early stages of the contract, while rates for certain finishing products are quoted very low, at cost, or even below cost to still make the overall price competitive. On the one hand, the retention monies will still be the same (as it is expressed as a percentage of the total contract price), however, where a contractor might run into financial difficulties and not be able to complete the project, or in the eventuality that a contract may be terminated for any reason, the contractor has walked away with all of his profit from the beginning of the contract. If the employer then tries to complete the project, he may not be able to get another contractor to complete the project at those low rates.

Six-star game lodge in Botswana

Noel Reid, formerly project manager and quantity surveyor for both Chief's Camp and Chobe Chilwero; employed by Abercrombie and Kent

Background

The Okavango Delta is a very large inland delta system located in the Kalahari Desert of Botswana, in southern Africa. It is a World Heritage property encompassing an area of 2,023,500 hectares with a buffer zone of 2,286,630 hectares. This unique wetland is fed by the Okavango River, which flows from the mountains of Angola. The crystal clear waters are home to some of the world's most endangered large mammals: cheetah, black rhinoceros, African wild dog and lion.

To facilitate tourism, the government of Botswana provide leased land to selected lodge operators for fifteen-year terms, upon which to build small exclusive lodges. The design and construction of the lodges are required to meet exacting criteria and comply with eco-friendly regulations. For example, the only concrete allowed is in the bund to the diesel storage area for the generator.

Building Chief's Camp

Having won a tender on a piece of land titled 'Chief's Island', and armed with a set of GPS coordinates, our professional team landed in Maun and transferred into two Land Cruisers, trailers attached and loaded with tents and provisions for the first site visit. The drive to the site was estimated to be five hours and, being the dry season, no boats were required to access the leased area. Arriving at the island in the late afternoon, the African sun was beginning to wane. Driving up onto the island through the long dry grass, we were struck by the beauty that only wild Africa can offer. In the centre of the island was a thorn tree and a small dry pan (pond), all was idyllic. We began to plan where we would pitch the tents for our two nights' visit, and spirits were high.

Our tranquility was suddenly interrupted when a very large male lion arose from the shade beneath the thorn tree. Lions, after all, are fundamentally large cats and very inquisitive. This area of the delta was seldom visited by humans and our intrusion into his kingdom was causing some concern. One lion in what was to be our campsite was less than ideal. When the heads of eight lionesses then appeared from the grass close by, we knew we were in challenging surroundings.

Undeterred, we chose a part of this small island for our campsite, and drove the vehicles around to flatten the grass. We parked up and unloaded the small pup tents and set out the camp. To prevent becoming an evening delicacy for the watching pride, a large fire was required, and to be kept burning all night. A large pile of wood was enthusiastically gathered by the team, albeit with one eye over their shoulders. With the fire lighting and the sun setting, many chilled beers helped to calm the nervous among us. Stories of Africa were the order of the day. Evening enveloped us with the sounds of nature, and a sky clear and filled with stars, resembling a chandelier; tired bodies began to relax.

Lions, it would seem, view the walls of a tent as impenetrable. While there is canvas between us and them, they do not disturb the occupant. But if the doors/flaps of the tent are open, the protection evaporates. The challenge arises when, having consumed numerous beers, that inevitable call of nature arrives in the early hours of the morning. There is only so long that the pressure and pain can be resisted. The question is how and where to go? Opening the flaps and walking to the edge of the camp when there is the heavy breathing of a lion close by is not a sensible option. Practicality prevailed; methods varied.

The early morning sunshine and much lion spore greeted us. Hot coffee was the order of the day and we began planning our setting-out works for the new twelve-bed lodge, public areas and back of house complex. The borehole drilling vehicle was due, and at least two lined boreholes are required with confirmed yields to support a lodge, ideally with a third in reserve. The 'chef du jour' prepared breakfast and the reality of morning ablutions arose among a certain few. It is very interesting to watch grown men, armed with spades and toilet rolls, stroll towards the nearest tree through long grass, making the 'distance choice' between privacy and safety: 'Where were the lions, how far shall I go?'

The borehole drilling team and equipment finally arrived. We began drilling in the ideal location as set on the masterplan. The first hole was drilled to forty metres: not a drop, just dust. We scratched our heads; standing in the middle of a delta, how could this be possible? A second hole was drilled in the same area; same result. Then Africa came to the rescue in the form of a giant leadwood tree. This species requires a good supply of water, so we drilled in close proximity to its knurled trunk with success and clear potable water flowed. Relief all around; with a second success, the site was now viable, the key risk removed.

The lions were, I am sure, watching our every move; however, in the heat of the day they were largely inactive and our work was not disturbed. In the late afternoon, they reappeared and seemed very interested in what we had been up to on their island. We retreated to the safety of the fireside, and gave them back their domain. The challenges of the night lay ahead.

We completed the lodge and infrastructure in a record six months, during which time the flood waters arrived and our island was surrounded by water. Many wonderful encounters with the local wildlife were had during the building process. Each morning at dawn, the architect and I would walk across the surrounding delta, a most magical place: unique in the world and a great privilege to experience. We watched the birds and the buck (no guns are allowed and it is every man/woman for themselves), but 'c'est la vie', live life to the full; it is, after all, not a dress rehearsal.

During the construction period, a low-level electric wire from the Land Cruiser was placed around the kitchen/meeting room area to keep the hyena out. Many evening meetings were disturbed by screeching as a would-be intruder was kept at bay. The practicalities of building in an area where the animals rule the roost beats an office in London any day.

One of the most memorable pictures from the construction album is the pride of lions encircling the seventy-foot high communications mast with the forty-strong site team up the mast, hanging on for dear life. Fortunately that day the lions lost interest, and moved on to capture easier prey. They were always around and, on some occasions, joined by other inquisitive cats. I recommend that you come and visit them. To hear them purring while resting against the wall of your tent is a very special closeness.

The investment by Abercrombie and Kent in Chief's Camp was a great success and led to a subsequent lodge development on the Chobe River in Botswana, named Chobe Chilwero. Do pay them both a visit and experience 'The Okavango and Chobe' for yourself; you will not be disappointed.

Guarantees/performance

The employer is entitled to require a performance security, sometimes referred to as a performance bond, which is a form of guarantee that the contractor will complete the works. If the contractor fails to complete the works, then the employer can call upon the performance security. Developers intending to build a hotel should discuss this with their potential lenders prior to setting the terms and conditions of the performance security, as often the lenders will stipulate in what form, and from which acceptable institutions, they would accept a performance security. This is normally a condition precedent in any loan agreement and therefore if the contract is signed prior to confirmation of all terms from the lenders, an employer might find himself having to renegotiate these terms with the contractor, who is not obliged to accept any proposed changes to the contract.

The performance security must be followed up on, because if there are any delays to the project programme, and there is an expiry date in the terms of the performance security, then the contractor must extend that expiry date in order to keep the security valid until completion of the project.

Notes

1 'FIDIC refers to the Client as the employer, whereas NEC refers to the Client, to avoid repetition below, the author will refer to the employer, using the FIDIC terminology'
2 In simpler terns, 'Yellow Book' is Design and Build, whilst 'Silver Book' is Engineering, Procurement, Construction/Turnkey contracts
3 Referred to a Project Manager under other forms of contract
4 Also known as overhead. This includes the contractor's overhead e.g. daily salary cost, fuel, equipment rental, phones, etc.

6 Disposal guide

This chapter deals with the disposal of a hotel or resort. It covers what is involved in the process of selling a hotel, how the sales price can be enhanced to ensure maximum value is realised, when is the best time to sell a hotel, how to choose the best agent and marketing strategies.

The process of selling a hotel

Hotels and resorts are usually sold as operational entities. As such, all operational agreements made with the previous owners are passed on to new owners to be honoured. This includes, for example, new bookings, equipment hire and staff contracts. This makes the sale of a hotel different from typical property transactions, where the property is sold without such complications.

Typically, staff and customers can get nervous if they know a hotel is being sold, especially if they find out secondhand, rather than from the owner themselves. As such, it is very important to work out how to address the issue of the sale process with the key customers and staff. If key staff leave, or bookings are not made, it will have a negative impact on the price obtained for the property, as well as on the earnings generated before the sale completes.

In addition, the hotel requires stock to function. This is usually audited on the day of handover, and a separate amount of money is paid to cover this, over and above the agreed sale price. This is not standard for property transactions, but is absolutely vital to both the buyer and the seller if the property is to continue trading as well as it should. If a fair price were not paid for the stock, then the seller would try to reduce the stock on site, which could lead to shortages and potential customer complaints. This would impact on trading for the buyer. As the income earned from the hotel remains with the seller until it is transferred, this lower income would also hurt the current owner as well.

It is also worth considering the impact of sale upon a franchised or managed property. Typically, there are clauses contained in various agreements that mean the transfer of these rights is not automatic, and the sale can be halted if the wrong person looks to buy the property. Alternatively, there may be a right for the management company to terminate the management contract and claim compensation in such an event.

Choosing an agent

Derek Baird, director of niche hospitality advisor practice Pacific Property Company

When buying a hotel, the owner likely gave careful consideration to the purchase and built up multiple professional relationships. Now it's time to sell and this time the owner has the choice of selling agent and has far more control of the process.

The choice is not simple. The appointed agent has to fully understand the owner's professional and financial circumstances, and must tailor a marketing process around these and prevailing economic circumstances.

When deciding on an agent, the private owner has to take into account that they will require a high level of direct personal contact throughout the process, and that both seller and agent need to be able to speak regularly and, crucially, openly. Thus the first decision is based upon personality and relationship. The owner needs to rely upon his advisor to be fully open with advice and feedback during the sale and, equally, the agent needs to have complete confidence that the client is keeping him fully briefed.

Trust and transparency are obvious qualities demanded from the agent, but dedication and time allocation are often overlooked. An agent working for a large, chain agency may have as many as twenty or thirty instructions at any point taking up his priority list which, when combined with the corporate demands for business development and internal company compliance issues, etc., minimise the time spent on client and process relationships.

Therefore, an independent agent is probably more likely to serve the direct needs of a private seller in terms of product focus and time allocation, and may also have more dynamic knowledge of the micro-environment of the hotel in question.

Multiple unit owners such as hotel groups or private-equity-backed entities may have a different priority in picking a selling agent. Existing relationships developed from building up the portfolio, combined with the need for a paper trail to demonstrate best value was obtained through the efforts of a nationally recognised agency, will in all likelihood lead them to appoint a well-known agency almost by default. Their likely intensive sell-side experience also leads them to require less input from the agent on routine matters.

The choice comes down to personality and process. A sales process can take weeks and will require multiple conversations with the chosen agent. It is key that both client and agent have a relationship that can allow for continued and regular communication to react to buyers' activities and demands.

Whether the seller is a third-generation family hotelier or a fifty-hotel corporate portfolio, the core decision is based upon the human aspect of the relationship. Then the sale can be undertaken to reflect this.

Agree what is required, fees and terms and conditions

The first step is for the owner and their agent to determine what is actually required, and then to agree fees and the terms and conditions of the instruction. It might be at this stage that it becomes apparent that a sale may not be in the best interests of the owner, or that a different approach may realise more value from the asset. It might, for example, become apparent that holding the property for a short while as changes are made to the business would allow the property to be sold for significantly more.

It is vital that the terms of the agreement between the owner and the agent are agreed and are very clear from the outset.

The agreement may be for sole or multiple agency, it may be an off-market transaction or a full marketed campaign, and in certain circumstances certain buyers may be excluded from the process. It is important that all parties agree at the outset if the process is to be relatively trouble-free and uncomplicated.

Undertake the initial due diligence to ensure the property will attract potential buyers

The due diligence work that is essential at the beginning of the sales process will include a full inspection of the property, so that 'the good, the bad and the ugly' of the property are known to the agent. Opportunities to enhance trade will be discovered, as will defects or potential future problems.

Any defects can then either be addressed, or potential solutions can be considered, so that the financial impact of such problems does not result in too low an offer.

The next step is to review the data, including details on the title (preferably a report on the title), statutory documentation (including planning records), financial records and employment data. Any licences, subleases or agreements with third-party operators will also be reviewed for their impact on the potential profitability, as well as looking at how these agreements may appeal or deter potential purchasers from buying the property.

Any treasure hunting opportunities should be sought and considered as to how they will best help sell the property.

Agree the sales process

The disposal method and process will be agreed, whether private tender, informal or formal tender of auction.

The specific marketing strategy, and potential buyers, will be discussed in detail, with a particular plan designed and followed by the parties.

One specific area that needs careful consideration is the choice of a 'guide price', which is outlined later in this chapter.

Private Treaty

The process of sale by private treaty is the method employed by most agents. It involves preparing descriptive details of the property and quoting a definitive asking price. Details are circulated and potential buyers may view the property, and either agree to buy at the asking price or submit an offer to purchase. Agreement to buy at this stage is subject to formal contracts being prepared between the vendor and the purchaser, and those contracts being signed and exchanged between the two parties. If several interested parties are introduced to the seller, those parties will be invited for 'best & final offer', thus ensuring the vendor receives the optimum price.

Pros
- Price is determined by the market
- Can generate higher sales prices than other methods
- Most buyers are comfortable with the process.

Cons
- Very reliant on a good agent
- No transparency in the transaction.

Auction

The property is advertised for sale by auction, rather than at a fixed price. Those interested in buying attend a competitive auction, conducted by an auctioneer, at which the person who bids highest buys the property. The successful bidder is legally bound to purchase when the auctioneer's hammer falls on his bid. He pays a 10 per cent deposit there and then and has to complete the purchase on the stated completion date – normally four weeks after the auction date. The buyer has to arrange finance and make any enquiries (including carrying out a survey) before he bids. It is too late afterwards. Property is usually only sold by auction when there is likely to be strong competition, negating the problems that some potential purchasers may not be able to attend the bidding process.

Pros
- Transparency of process (it can be demonstrated that the best price was acheived)
- It appeals to cash buyers
- Speeds up the sale process
- It is good for properties with serious defects, where buyers might pull out
- Where the sale price cannot be predicted, this can help ensure the property is not sold too cheaply.

Cons
- Potentially difficult for potential buyers to attend, so may lower the competition for the property
- Does not always generate prices comparable to private treaty
- Very dependent on the quality of the agent.

Informal Tender

Sale by informal tender is a process where written offers will be invited (sealed bids) and a closing date for such offers published. All offers are opened at the same time. Generally, the vendor is not committed to accepting the highest or any offer. The offer is not binding and on acceptance of any offer, the transaction proceeds subject to contract.

Pros
- Transparency of process (it can be demonstrated that the best price was achieved)
- Sets up a fixed timeline for completion of the process
- Good for properties with major defects.

Cons
- Unusual method of disposal which can deter those unfamiliar with the requirements of the process
- Does not always generate prices comparable to private treaty
- Very dependent on the quality of the agent.

Formal Tender

When a property is sold by formal tender, as with an informal tender, the sale will be advertised with a deadline by which prospective purchasers must submit their bid. Each tender document from the bidders must include the full legal contract for sale, and all bids have to include a banker's draft as a deposit on the contract. The bids are opened by the vendor or agent (representative). As soon as the 'best bid' is selected, the banker's draft is accepted and contracts are automatically exchanged. The successful bidder is then committed to the contract and will have to complete the sale on the appointed date. If the successful bidder fails to complete the sale, they will forfeit their deposit and further costs may be incurred. Generally rarely used due to its complexity.

Pros
- Transparency of process (it can be demonstrated that the best price was achieved)
- It tends to be used mostly for land disposals, where the formal use is as important as the price raised.

Cons
- Does not always generate prices comparable to private treaty
- Very dependent on the quality of the agent.

Figure 6.1 Methods of disposal

Preparation of the supporting materials

This involves the production of any marketing collateral (including brochures and advertisements) and preparing the 'data room'. A data room is typically an online area where access can be granted to specific parties so they can fully evaluate the opportunity, while still retaining a level of confidentiality for the seller. Typically, it will include legal documentation (leases, title documentation, operating agreements, statutory licences and consents, for example), financial reports (with full historic and budget management accounts), a property's details (floor plans, surveys, reports, valuations, photographs), general market information (data on the local area and local market) and employment sector details (staff details and sample contracts).

Marketing

The next stage of the sale process is to run the marketing campaign along the agreed lines, making any changes to the initial strategy as are required to help ensure the highest and best price is achieved.

Ideally, the marketing campaign will generate strong interest from at least three or four parties, resulting in a bidding war, to drive the price above the initial offers received.

Review offers and determining the most suitable buyer

The next step is to review all offers and determine which offer is the most favourable to the seller. This may be the highest offer, the quickest or the offer that has the least risk (an unconditional offer, or an offer from a reliable source).

Certain buyers have a reputation for offering the highest bid and then at the due diligence stage reducing their bid to much lower levels. The risk of 'chipping', as it is sometimes known, can lead to certain buyers' offers being excluded from the process at the outset.

Completing the transfer

Typically, the seller will provide an exclusivity period (if not entering a contract race) to the purchaser, while they work through the due diligence for financial, legal, valuation and condition. A contract race is when an agreement is made to sell a property to whichever potential buyer can exchange and complete the transaction quickest. This will involve more than one party spending significant amounts of money on due diligence, and is only usually possible when a property is particularly desirable to investors, or when the seller offers to reimburse the losing parties of their costs if another party completes ahead of them.

The next stage is to exchange contracts with an agreed completion date (and deposit paid). There will usually be a stock audit on the day of handover, with an additional sum paid on the basis of actual stock, prior to the completion of the transfer.

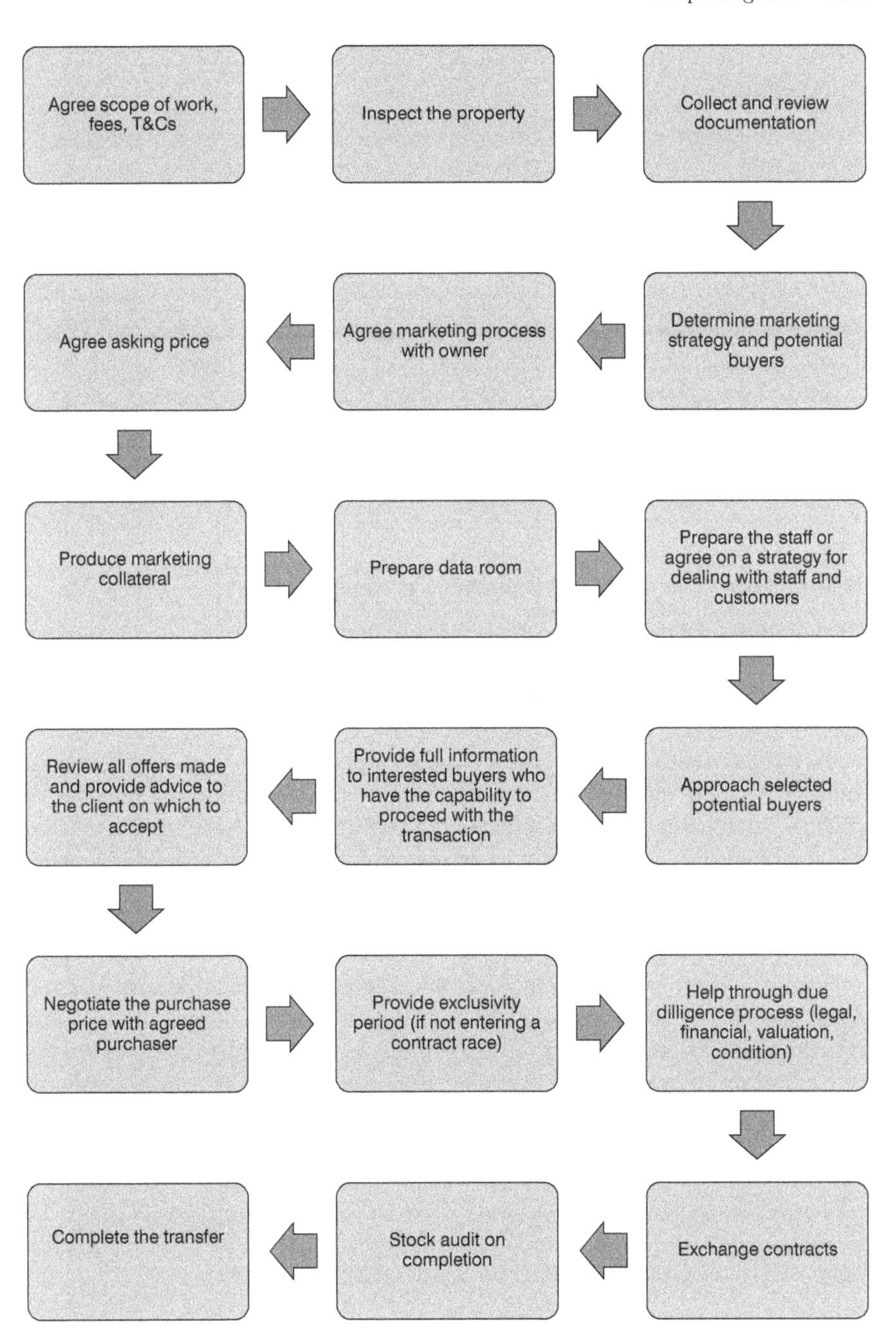

Figure 6.2 Process of a hotel sale from an agent's perspective

Alternative uses

One key thing a valuer needs to be aware of is a higher alternative-use value. This happens more often than expected, especially when a hotel is no longer ideally suited to the local market requirements. I was involved in the purchase of a tourist-class hotel in London during 2006 that cost £3.1 million as an operational unit.

Over the next two years, planning consent was granted for residential use, and nine flats were created, selling for an average of £1.45 million. The overall value released from the scheme was £5.05 million after the purchase price, development costs and finance costs were taken into account.

It is arguable that the vendor should have been paid more than the £3.1 million that they received, and had they considered the alternative use, may have held out for a higher sale price.

How to ensure the hotel is as valuable as it can be

It is important for anyone involved in the hotel business to have an understanding of the various factors that will determine how much money their hotel could sell for, as this will inform them of the optimum time to try and sell the property, or when the sale price is likely to be at its highest.

As discussed earlier in the book, three key factors come into play when assessing value. First, how much profit does the hotel make? Second, what multiple of profit is a purchaser prepared to pay for the type of hotel? Third, what condition is it in? Many other factors affect these key areas but, for the purposes of looking at how to get the highest price when selling your hotel or resort, it is important to focus on the bigger picture in the first instance. Anything that positively impacts these three factors will help enhance the potential sale price.

There is one outlier here that needs to be considered. Although a hotel/resort is sold as a trading entity, if the property has a higher alternative-use value (for example, if it were converted to a care home for the elderly, or into private residential dwellings), then a purchaser may consider paying a higher value if they believe a change of use is actually possible and potentially profitable. It is not always easy to release this 'additional value', so rarely will the full excess value (or 'hope value') be paid by a potential purchaser. However, if the seller has taken steps to open up the opportunity, for example, by gaining planning consent for a change of use and undertaking an independent feasibility study, then the offers received are more likely to reflect this additional value.

Potential new supply

Any new hotels could have an impact on the trading potential of a hotel, if demand for hotel rooms (and facilities) does not increase at the same pace as the new supply. In Luanda, the capital of Angola, between 2010 and 2015, the supply of quality bedroom accommodation went up by 120 per cent, which had a detrimental impact on RevPAR, as demand for rooms did not increase by the same amount.

Existing supply upgrading

If all the competitive hotels are enhancing their 'product' while the subject hotel is not, this is likely to lead to the hotel being seen as less desirable by the market, with a possible impact on the volume of room nights sold, the rate achieved or potentially on both.

Existing supply closing down

When supply decreases, as long as demand for hotel rooms does not decrease at the same rate, it is likely that performance (across the whole market) will improve, allowing a well-placed, well-managed hotel to see business increase.

Changes in local demand

If a large local business closes down, its requirement for bedroom nights, meeting facilities or restaurant bookings ceases, which can have an adverse impact on trading potential. On the converse side of the equation, if new businesses open up, or expand, taking on new staff, this could well lead to enhanced business.

Changes in demand for all parts of the property

Whether looking at changes in the supply and demand equation at the health club (in terms of membership numbers or membership rates), the spa, golf course, bar or restaurant, any major changes that impact 'the satisfied demand' at the property can have an impact on trading potential.

Changes in dynamics of the local demand profile

A new road or railway line linking the property's location to potential new demand is likely to have a positive impact on the trading potential of the business. New airport links are of vital importance for the growth of hotel demand in new destinations.

Ryanair almost single-handedly enhanced demand to a wide range of secondary and tertiary leisure destinations across Europe by opening up cheap transportation links with potential source markets. However, changes in infrastructure that deter people from staying, or lessen the need for them to stay, for example, the construction of a bypass negating the need to stay overnight at a destination, will potentially have a negative impact on trading.

Changes in cost structure

A rise in energy prices will potentially have a direct impact on the profitability of a hotel. Likewise, increases in staff wages and benefits could also impact directly on hotel profitability.

Changes in supply chain

If food supplies suddenly become scarce, costs can increase, leading to reduced profitability. However, if new suppliers become available, potentially offering well-priced local supplies that reduce the need for transportation costs, this can have a positive impact on profitability.

The multiplication factor applied to profits is a significant factor in maximising the value realised from an asset sale. Many factors impact what the potential buyer will determine is the appropriate multiplication factor, as listed below.

Price of potential alternative investments

If the typical purchaser of a boutique hotel would alternatively consider investing in fine wines, and fine wines are currently generating 15 per cent returns, then this will have an adverse impact on the demand for 10 per cent returns on hotel investments. Typically, most investors have alternative investment options and if they are seen as too expensive, then hotels will look desirable, while if they are shown to produce better returns, then hotels may appear less desirable.

Supply of similar investments

If there is a shortage of 'investment product' and a large amount of competition for each investment opportunity, this will generally enhance the multiplication factor applied to a hotel investment. This 'wall of money' looking for a home will compete with itself until the price becomes uneconomic. However, when there is an over-supply of 'investment product', potential investors will be able to pick and choose which investments to buy, typically leading to lower prices.

Prevailing economic climate

If confidence in the economy is strong, then investment yields tend to be stronger, reflecting that confidence. When people are concerned about the economic climate, it can lead to lower prices, or just fewer deals overall, as more buyers decide to delay making new investments.

Prevailing investment climate

The prevailing investment climate is linked to the general economic climate, but is more specific to the investment subset. If people fear that the hotel market has overheated, or that demand may be lower in the future, it generally has an adverse impact on yields. The same is true in reverse; if people feel there will be enhanced demand in one market, for example, feeling that hotels in New York are too cheap, then yields will tend to sharpen.

Amount of supply in the direct marketplace

If there is a shortage of luxury hotels in Monaco, then investors may well be prepared to pay a higher price to buy one. Conversely, if it is considered that there are too many mid-market resort hotels in Mombasa, prices are likely to decline.

How well-suited the property is to the specific investor's requirements

Specific purchasers will determine the yield that a property is worth to them, depending on how well it suits their needs. An international hotel chain that is missing a hotel in a key city like Tokyo may well be prepared to pay more (whether buying the asset, offering more rent for a lease or accepting lower management fees) to fill that gap in their inventory, for the benefit of the whole chain.

The quality of the tenure

The nature of the tenure has an impact on the value of an asset. Typically, freehold title is most valuable, with leasehold title generally lower in value (all things remaining equal), with licences or customary title less valuable still. In essence, anything that lowers the control an owner will have on the land and lowers their ability to sell the property, the lower its value.

Potential development of the property in the future

If the property has any latent value, for example, the ability to develop more rooms, private residential apartments or condos, or to redevelop the whole site for a more valuable use, then this 'hope value' may well factor into the price a person might be prepared to pay for the asset.

Skill and financial standing of existing operator

If the hotel or resort is being sold as an investment (i.e. subject to an operational lease or a management contract) then the perceived quality and perceived financial standing of the operator will impact on the yield investors are prepared to pay.

The nature of any operating agreement

The lower the risk to the investor, the higher yield they are usually prepared to pay for an asset. As such, a fixed-stream rental income where the rent is easily affordable will attract a sharper yield than a management contract where all the operational risk is passed on to the investor.

Reputation of the property

Market perception of a property, especially if it is perceived to be a trophy asset, will have an impact on the yield that investors are prepared to pay for a hotel. The condition of the asset has an impact on the price that will be paid, though slightly less directly than do the other two key factors. If a property requires capital expenditure to bring it up into optimum condition, the cost of such works are obviously an important consideration. However, a property in poor condition can also generate lower income for the owner (impacting on profits) and be less desirable to investors (impacting on the multiplication factor applied to the investment).

Resolving title issues

Reviewing the title and dealing with any issues that become apparent, whether resolving them or preparing for them during the due diligence process, is an important step in enhancing the sales price.

Ideally the hotel will be freehold, with good and marketable title, with no encumbrances and restrictions at all upon the use of the property. If that is the case, then the sales price will be as high as the trading allows.

If the property is leasehold, the agreed rent will obviously have an impact on the profitability of the hotel, and restrictions on use could have profit implications (not selling alcohol, for example).

However, legal issues tend to impact the multiplication factor most, with lower multiplication factors usually applied to anything that restricts value. In addition, the term of the lease has a direct impact on the value of an asset, as the value is dependent upon the term it is held for.

Therefore, if a leasehold term is nearing expiration, agreeing a new term could have an immediate positive impact on the sale price that is achieved.

If there are encumbrances and restrictions upon the title, and if these can be removed, or the impact of them lessened, then this will enhance the potential sales price.

In the event that issues cannot be resolved, for example, a restrictive covenant forbidding the use of the property as a hotel that cannot be removed, then looking into an insurance policy might be the best way forward to mitigate the issue.

What is a trophy hotel?

The hotel market is one where certain classes of hotel command prices where a reasonable return on capital seems unlikely. Trophy hotels, where the level of pride of ownership is so high, tend to be purchased at prices almost irrespective of the trading potential of the property.

There is no single definition of what constitutes a trophy hotel, though it tends to be an upper-upscale hotel at the top of its local market, usually with a profound sense of history. The key to whether a property is a 'trophy' is determining to what level potential owners will go to own that particular property.

In certain markets there are a number of trophy assets, for example, in London these would include The Ritz, Claridge's, The Savoy, Mandarin Oriental Hyde Park and The Dorchester, while in Paris these would include Le Bristol, Le Meurice, Hôtel de Crillon, Plaza Athénée, George V, Le Vendôme and Le Royal Monceau. In such cities, there are other properties that could be argued to be trophy hotels based on the quality of the accommodation (for example, in London, these might include 45 Park Lane, the Four Seasons, The Lanesborough, Brown's, The Connaught or the Bulgari), but the test is always: who is interested in the assets and how much will they bid?

The problem for the valuer on assessing the value of a trophy asset is that it really is difficult to be exact as it can depend entirely on the appetite of the market at the date of sale. In addition, a large number of deals are done off-market, so the full interest of the market is often untested.

Trophy hotel – examples

Figure 6.3 Peninsula Hong Kong

Peninsula Hong Kong

The Peninsula Hong Kong is the flagship property of the the Peninsula Hotel Group. It opened in 1928, founded by members of the Kadoorie family, built with the idea that it would be 'the finest hotel east of Suez'. It was famed for its large fleet of Rolls-Royces painted the distinctive 'Peninsula green', which were used to transport guests to and from the hotel.

The hotel has great historical significance, and on 25 December 1941, at the end of the Battle of Hong Kong, British colonial officials led by the Governor of Hong Kong, Sir Mark Aitchison Young, surrendered in person at the Japanese headquarters on the third floor of the hotel.

Raffles Singapore

Raffles Hotel Singapore opened in 1887 and it is arguably one of the few remaining great nineteenth-century hotels in the world. A century after its opening, the hotel was declared a national monument by the Singapore government, after which it underwent a complete restoration at a cost of US$160m. Over the years, some of the most famous personalities have stayed at the hotel, including writers such as Somerset Maugham, Rudyard Kipling and Ernest Hemingway, and film director Alfred Hitchcock.

Figure 6.4 Raffles Singapore

The property was last purchased in 2010 for a reported price of US$275m, equating to US$2.7m per bedroom, which was at that time the highest price ever paid for a hotel on a per bedroom basis.

Choice of management

The choice of management is of vital importance to both potential earnings and to the multiplication factor applied to the investment. Having the optimum operator for a property will enhance not only the hotel or resort's revenue potential, but also how efficiently that income is converted to earnings for the investor.

Appointing an international operator to manage a hotel can have a significantly positive impact on the returns the owner receives from the investment.

International operators can often improve performance over and above that generated by an independent hotel operator, increasing revenue production while also using their size to generate cost savings, thereby improving the owner's return. In addition, when the hotel is finally sold, it usually attracts a higher multiple of earnings if it is internationally branded (and therefore a higher price).

However, if it is 'the wrong brand', or in the 'wrong structure' (a poorly conceived operating agreement for the specific property type), it could have a negative impact on operational earnings potential and therefore the capital value of the asset.

Having the correct management team in place is important for vacant possession properties as well. Having an experienced, enthusiastic team in place is of real benefit to the operation of the hotel or resort, often generating better returns. Also, such enthusiasm is usually noticed by potential buyers, to whom the property will appear more attractive.

The difference that the choice of the management team can make to the potential sales price is much larger than might be expected. Even though hotels are usually sold with the potential to change the management structure, the actual trading and projected trading is very important to the price that potential purchasers will pay for the asset.

Staffing

The quality of management leads straight through to the quality of the team below them. If the property has the optimum number of staff employed, an experienced team in place and the appropriate flexibility in place, then the price paid should not be adversely impacted by staffing issues.

The amount of staff varies around the world, as well as depending on the type of property. Super-economy hotels have as little as one member of staff per ten rooms, while a super-deluxe safari lodge can have as many as four staff members per room. The level of service, the amount of additional facilities offered over

Hire for passion, not just talent

Petra Devereux – Elevation Personnel

It has been said to 'hire for passion, not just talent', and I could not agree more. Over the past six years, passionately heading-up my own talent acquisition firm (to a large degree in the hospitality sector on the African continent), the basis upon which the exceptional hotel general managers stand out, and take their hotels to the next level of profitability and guest satisfaction, stems from their absolute passion for their job.

Talent most certainly plays a crucial role in the ability of a general manager to skilfully analyse sales figures, astutely monitor guest-satisfaction reports or to direct staff. However much passion drives sales and revenue by integrally participating in the sales processes, strategies and execution, it also leads by example and exudes 'presence' with guests and staff alike. In so doing, it motivates, communicates, rewards and drives a hotel team to operate on a higher level. Such passion also ensures that a hotel stays abreast of global trends. In these instances, staff are generally proud participants in such passionate leadership and as such, they become passionate themselves. Long hours, hard work and ensuring optimal experience for guests are a requirement for this industry, and talent alone does not make for a successful hotel general manager. Passion is key.

and above basic rooms, and the quality and skill of the available labour pool all impact on staffing requirements.

Having too many staff, or indeed too few staff, will impact upon trading potential, so it is important to review the staffing levels to ensure they are optimal.

Typically, the staff will be allocated to certain departments, where their training will ensure a consistent standard of service. There has been a trend towards multi-training, allowing flexibility within staffing rotas, though this does require good training and excellent staff.

Timing of the sale

Taking into account the three key factors that impact on value mentioned above, each one will potentially have an impact on the optimum time to sell. The ideal scenario is finding a time when trading is as high as it will get (in present values), when demand for this type of asset is at its highest and when the condition of the hotel is excellent.

Determining when profit is at its highest in present values is dependent upon knowing the market's cycle. Just for clarity, this refers to optimum trading in present values, as if we take inflation into account, a stable hotel will continue to enhance its monetary profits each year, without actually enhancing the actual profit. If growth in profit is higher than inflation, then the trading of the hotel is not stable.

A typical new hotel takes a number of years to reach stable trading levels. A new development will typically take between three and five years to reach stable trading after opening, and once trading has stabilised, that will usually be the best time for a developer to sell, if all other factors stay the same.

New supply into a market, or changes in demand (for example, a new office park opening up, new airline routes, a new bypass taking business away or main customers closing down) can have a major impact in future trading, and should be factored into any timing considerations for when to sell an asset from a profits perspective.

Preparing for sale

A wise investor plans for the time to sell the hotel from the very beginning, at the development phase or even during the initial process of an operating hotel. If the investor knows what they want to do in the first instance, they can ensure that it is built into their planning as long-term holds have different strategies than medium-term holds or early disposals.

It is essential that all the due diligence paperwork is collated by the owner and agent before going to market. Nothing stops a very keen potential buyer being interested in a transaction than having to wait for the paperwork to review. It is imperative to the success of a transaction to be able to follow up such interest immediately, to build the momentum in the transaction so that it can be pushed through to a successful conclusion.

Choice of agent

When disposing of the hotel, an experienced specialist hotel agent will usually be best placed to secure a quicker sale at the best possible price. The chosen agent must be competent in the specific location, with the type of property and with the type of structure being sold.

Generally, the best agents will only work on a sole agency basis, to ensure they can fully control the disposal process, thereby enhancing the chances of a successful transaction.

Sometimes it appears prudent to appoint multiple agents. The logic behind this is that two sets of agents will approach twice as many potential purchasers, twice as quickly, making the chance of a successful sale twice as likely. However, this is rarely the case. The problem with joint-agency instructions is that no single person is responsible for the whole disposal program, so certain buyers may not be contacted through accidentally being forgotten, or joint agents may wish to keep certain 'buyers' away from the other agents, and so do not contact them.

Treasure hunting

Ideally, all treasure hunting opportunities will have been explored before the property is brought to market. They may already have been constructed, in which case full value should be sought for them. They may just have been developed as a concept, with drawings, planning consents, a feasibility study and development costs in place. In that case, the potential buyer may pay a higher price to take this potential into account.

If the assessment of the work (costs, feasibility study, design) has not been carried out, and planning consent is not in place, then little value will be attributed to such potential. However, it is still worth bringing it to the attention of potential purchasers as it might be a deciding factor on whether to bid for the property.

Marketing

It is essential that the property is marketed to the correct people. There is no point, for example, in advertising a property to international buyers when non-locals cannot own property in the country in question. In addition, certain types of properties and structures attract certain types of buyers, and it is a waste of time to contact investors who have no interest in the type of property that is being offered for sale.

There are times when it is prudent to speak informally to one or two buyers in an off-market scenario to best complete a sale. A competent agent will advise on this, on a case-by-case basis. However, in most situations, the desire for secrecy can limit disposal options, usually reducing the potential buyer pool, and possibly even lowering the price attained at sale.

Sometimes there are advantages in placing advertisements into the press to attract potential buyers, especially when the likely buyer is an unknown quantity (for example, a 'lifestyle' type of property).

Appropriate guide price

Research has been carried out that suggests that when a property is marketed with a realistic guide price, it attains a higher sale price than when an inflated asking price is used instead. The skill of working out what is an appropriate guide price is a simple valuation issue.

If the property is priced too expensively, it may deter potential investors from looking at it. If a property is worth €50m, then a guide price of €70m is not suitable, in the same way that €30m is also not suitable. If the price is much higher than the true value, then potential purchasers will feel they are wasting their time looking at the property. If someone is asking €70m, most buyers would assume that the seller would be unwilling to discount the sale price by 29 per cent. Having too low an asking price will attract many potential purchasers, but these are likely to be attracted purely by 'a bargain', and will lose interest when the property starts to be priced appropriately.

There is also research that shows that there is such a thing as an 'anchoring bias' and that presenting a low asking price sets that number as a fixture in the

The anchoring bias

Matthew Harper, researcher – Leisure Property Services

The 'anchoring bias' is a cognitive bias which causes people to latch onto the first piece of information they receive when making decisions. For instance, when you hear that the hotel down the road just sold for US$20m, this one looks like a bargain at US$12m, despite its actual value. Even as you learn more about the hotel and why it has a lower price, you are still anchored at US$20m and continue to think that the hotel at the lower price is a good deal, despite its value being lower than the other.

However, the anchoring bias only works up to a certain point. Once the anchor becomes excessively high, a potential buyer is unlikely to become tied to it, making it useless because they know that the initial price is too high, so they ignore it.

When selling, it is important to remember to keep the anchor higher than the actual value, but not so high that the potential buyer sees it as preposterous and doesn't become tied to it during negotiations.

While buying, you need to be aware that the seller is probably aware of the anchoring bias and will probably be trying to use it to their advantage, so you need to make sure that you are not anchored by an irrelevant factor, and instead focus on more important bits of information.

mind of potential buyers. In this instance, the higher the price, the better for the seller. However, in markets where most buyers use expert valuers to carry out the due diligence, this 'anchoring bias' is rendered redundant.

Timing of the sale

Timing of the marketing campaign is important, and not just in terms of what might impact on the trading or yield selection applied to the hotel or resort. Timing of the campaign, in terms of the time of the year, can be important. Avoiding key holiday periods is also important, because if the marketing starts when potential buyers are absent (say August for a hotel in France that would appeal to French buyers), it might lead to them either completely miss the opportunity, or assume they are too far behind and therefore not look at it seriously.

Sending out marketing particulars on a Friday in Dubai may not also be the best plan, as ideally you want the option to purchase to be very visible to the potential buyer, not buried under a weekend of email traffic.

When selling a seasonal hotel, it is generally best to market it while it is open, so that potential purchasers can visit it when it is at its best.

Enhancing demand

This is what every good agent strives to do, and it can separate a good agent from a great agent. The key ability is to present a property to the right people in the right way, and to ensure that demand for the property is as high as it can possibly be.

The key is working out how potential buyers want to receive their opportunities, on what day they are most likely to have the time to consider opportunities (or just as importantly, when the really bad times are when they have no time to review new opportunities), as well as working out who the most likely buyers for an asset are, and how to attract other potential buyers to the opportunity.

It is crucial to have a very experienced agent, one who knows the market (for the particular type of hotel or resort) very well. Using a local agent when international buyers are the most likely buyers is not the best choice. The same is true in reverse: when the buyer will be a local buyer, you need an agent that can tap into the local market easily.

Part II
The valuation process

The second part of the book outlines the valuation process, providing detailed examples and highlighting to investors some of the issues that need to be considered.

It discusses the valuation process for hotels and resorts specifically, though all the key points are equally valid for all types of leisure properties. It outlines the key methods of valuation, providing simplistic examples.

The importance of due diligence is discussed, outlining the various types of due diligence including legal, statutory and condition.

The chapter on financial due diligence explains the importance of understanding the business. It provides a number of worked examples on how to analyse the operation, so accurate projections for future performance can be made.

Valuation due diligence is addressed, highlighting the inspection process, assessing the local hotel market, the interview with the manager and finally, how to analyse all the data.

7 Methods of valuation

Introduction

There are five key methods of valuation which are used to value real estate, outlined below. It is important to note, however, that all of these rely heavily upon comparable evidence, without which any of the methodologies would provide nothing more than a mathematical exercise.

It is likely that the hotel valuer will have to work at some time with all five of these methods, and a thorough understanding of the methodology of each type of valuation involved is important.

1: The comparison method

This method requires the valuer to look at comparable sales/lettings and then apply a value to the subject property by adjusting the evidence from similar transactions to meet the criteria of the subject property.

This is most commonly used for properties such as houses or flats sold with vacant possession and is usually based on a price-per-property (e.g. two-bedroom, one-reception properties in the block of flats sell for £650,000) or price-per-square foot (houses in the location sell for £2,900 per square foot).

Evidence of the sale prices of residential properties is usually freely available and can be analysed quite effectively, albeit that judgement is still required by the valuer to determine the relative advantages and disadvantages of the subject property and the relevant comparables.

In terms of hotels, the comparison method is most commonly used when the property itself, rather than the income it can generate, is the primary driver behind the purchase, for example, for 'lifestyle hotels' or some trophy hotels.

Example: Hotel Boon

The Hotel Boon in Japan is being valued on a specific date. It is a forty-bedroom, tourist-class hotel in a market where most tourist-class hotels are purchased on a per bedroom basis.

Table 7.1 Comparable transactions

Hotel	*Size*	*Date of sale*	*Price (JPY)*	*Price/room*
Hotel Harmison	50	3 months before	900,000,000	18,000,000
Hotel Hoggard	30	2 months before	540,000,000	18,000,000
Hotel Flintoff	45	1 month before	810,000,000	18,000,000
Hotel Giles	50	1 month before	918,000,000	18,360,000
Hotel Jones	42	2 weeks before	756,000,000	18,000,000

There has been significant transaction activity recently, with the most comparable transactions listed in Table 7.1

The evidence suggests the likely comparable value is between JPY 18 m and JPY 18.36 m per bedroom, with most transactions showing a sale price of JPY 18 m per bedroom. As such, the value of the Hotel Boon in this instance is JPY 720 m.

Standing back and reviewing the valuation, the value represents a 12 per cent yield on EBITDA on the last year's accounts, which is appropriate for this particular market.

2: The investment method

The valuer assumes that the income generated on an annual basis from the property is sustainable in the future (or makes whatever allowances are appropriate), and then applies a relevant multiplier to this income stream, and deducts transaction costs.

The investment method of valuation is probably the most commonly used method of valuation for commercial premises, as the majority of retail, office or industrial transactions occur with the property sold when occupied by a tenant.

As such, offices with vacant possession can sometimes be significantly less valuable when compared to leased-out properties. It is worth mentioning here that a hotel with vacant possession is not like an office with vacant possession. Hotels are sold as trading entities and as such, 'with vacant possession' does not mean vacant or empty, it means without an operating agreement (lease or management contract), but still with all the bookings, staff and other items required to operate the business so that the hotel will be generating income immediately.

The investment method has become more important in the hotel world since the mid 1990s, as more properties are operated on leases and/or management contracts.

The investment method is either carried out in present values (known as the income capitalisation method) when a capitalisation rate is applied to the income, or in future values (known as the discounted cashflow method – DCF). The DCF method utilises a discount rate rather than a capitalisation rate to value the income stream.

Table 7.2 New lettings

Hotel	*Tenant*	*Bedrooms*	*Date*	*Term*	*Rent (£)*	*Rent/rm (3)*
Hotel Randall	Travelodge	110	3 months before	15 years FRI terms	715,000	6,500
Hotel Botham	Travelodge	100	2 months before	15 years FRI terms	675,000	6,750
Hotel Gooch	Travelodge	90	2 months before	15 years FRI terms	630,000	7,000
Hotel Boycott	Travelodge	105	1 month before	20 years FRI terms	729,750	6,950

Example: Hotel Gower

This example is a budget hotel, let to a well-known hotel chain on an FRI basis, in provincial UK. The market for such properties is dominated by purchasers who are looking for simple investments, buying such properties for the rental income they provide. As such, most purchasers look at such properties almost exclusively using the investment method, and then cross-checking, comparing the end value with the value on a trading basis (profits method).

Hotel Gower is a 100-bedroom hotel let to Travelodge on a fifteen-year FRI lease, on a shell and core basis. The rent received is £700,000 p.a., with rent reviews every five years, linked to RPI increases.

In this example, there is a selection of comparable hotel transactions, relevant to the subject hotel. Each is relatively similar in size, let on similar terms to the same tenant and located in comparable locations (in terms of trading potential and investor demand) (Table 7.2).

The rental levels range from £6,500–£7,000 per bedroom, which suggests the rent passing on Hotel Gower (£7,000 per bedroom) is in line with market levels, and so should be sustainable (Table 7.3).

Table 7.3 Hotel transactions

Hotel	*Bedrooms*	*Date*	*Price (£)*	*Price / room (£)*	*Rent (£)*	*Yield*
Hotel Strauss	100	4 months before	10,450,000	104,500	700,000	6.33%
Hotel Hussain	92	3 months before	9,400,000	102,174	653,200	6.57%
Hotel Trescothick	110	3 months before	10,500,000	95,455	742,500	6.69%
Hotel Prior	100	1 month before	10,000,000	100,000	685,000	6.48%

Note: The tenant in all cases is Travelodge

The evidence suggests that a net initial yield (NIY) of around 6.5 per cent is appropriate. As such, the value of the hotel is £10.2m.

Standing back and reviewing the value on a per-room basis, it breaks back to a £102,000 per bedroom, which is deemed appropriate for such an investment property.

Then, looking at the property on a trading basis, it is estimated that it could generate an EBITDA of £1m per annum, and a capitalisation rate of 10 per cent would be appropriate, suggesting that if the lease were forfeited, the operational value would be in the region of £10m. As the difference between the investment value and the operational value was relatively similar, there is no need to adjust the original valuation.

3: *The profits method*

The profits method of valuation requires that the valuer looks at the potential profitability of a property and applies the appropriate multiplier to the net profit (EBITDA) to assess its value.

The profits method is used when purchasers are attracted to the specific property based upon the profitability of the unit, and a simple comparison with other similar transactions will not provide an accurate enough valuation.

It is the method most commonly used to value trading properties and not just hotels. However, it is not used in isolation from the market. The assessment of the potential trading will be based upon the valuer's knowledge of how comparable properties trade, and the multipliers applied to the EBITDA will be based upon comparable evidence. When undertaking the valuation, the valuer has to apply significant amounts of judgement to complete the valuation.

Example: Hotel Tendulkar

After a detailed analysis of the resort's trading potential, it is determined that its sustainable EBITDA is INR 1,105m per annum, from a turnover of INR 3,200m.

Evidence shows that similar properties sold at a 9 per cent yield, suggesting the gross value for the resort is INR 12,277m.

Standing back and reviewing the result, the value breaks back to INR 30.7m per room, which is deemed appropriate for a property of this type.

Table 7.4 Stabilised year

Turnover (INR)	3,200,000,000
IBFC	1,405,000,000
%	43.91%
EBITDA	1,105,000,000
%	34.53%

Figure 7.1 The profits method

4: The residual method

The valuer calculates the end value of the completed development (the gross development value or GDP), and then deducts all the necessary costs of construction and finance, including the appropriate level of developer's profit, to calculate the resulting site value.

The residual method of valuation is used to assess site values where there is insufficient comparable evidence of site values, and where developers buy the site based on what the end development will be worth, minus the cost of developing the property.

In terms of hotels, many sites are purchased based upon prices that are calculated with reference to the end value of the completed development, because comparable site evidence usually requires a lot of analysis before it is truly useful.

However, the method is still exposed to comparable evidence. The building costs are all market tested; even if the developer has agreed a turnkey development package at a fixed price, the valuer will analyse this against 'average market costs'. In the same way, finance charges, contingency levels and levels of developer's profits will be linked to normal market parameters.

The end result will then be reviewed and any alterations that are necessary will be made to ensure the end value is in line with market parameters.

Example: Hotel Lara

In this example, a piece of land is being valued that has consent for a 400-bedroom hotel resort, with a golf course, spa, conference centre and 100 luxury residences.

Figure 7.2 The residual method

The gross development value (GDV) has been calculated at AUD 200m, taking into account the phasing of the sales of the residential units.

The total development costs, including building costs, professional fees, contingency allowances and finance costs are estimated to be AUD 130m.

The developer's profit for such a scheme is deemed to be 20 per cent of the GDV.

As such, the residual amount available to buy the land is AUD 30m.

Standing back and analysing this, the value equates to AUD 1m per acre, and approximately AUD 57,000 per bedroom, after excluding the residential units from the scheme, which seems appropriate.

5: *The contractor's method*

In this specialist method of valuation, the valuer looks at the cost of replacing the property, and assesses its value after reducing the rebuilding cost by an appropriate factor for depreciation and obsolescence.

The contractor's method of valuation would be used for properties such as public toilets, power stations or other specialist properties where the previous methodologies could not be used.

It is sometimes used for hotels when no market exists for such properties. However, the requirement to use this valuation methodology for hotels is reducing as the market in hotels is expanding worldwide, and other methods should always be used ahead of the contractor's method, if at all possible.

Example: Hotel Donald

This example is for a five-star hotel that was built by the government solely for the purpose of entertaining foreign dignitaries. It is not open to the public and has been built in a location that dictates that it would not be of interest to potential hotel operators or investors.

As such, the comparison method, the investment method and the profits method of valuation cannot be used to accurately assess the value of the property. As a result, the contractor's method is deemed the appropriate method for the valuation.

The estimated replacement cost of such a property is KES 3,750m. This is based on current building costs to create a similar property, in style of and size.

The level of depreciation and obsolescence is deemed to be 25 per cent.

As such, the value of the property is deemed to be KES 2,812.5m.

Hotel valuations

When carrying out the market valuation to determine the value of the property, the valuer is trying to reflect what the buyers currently in the market would pay for a property in the marketplace at that time.

The best way for any valuer to determine what the market will pay is to be thoroughly immersed in the market, and to adopt a valuation methodology that best reflects the approach taken by the active purchasers, for that type of property at that time.

It can be very difficult for a valuer to assess the value of a hotel if they are kept in isolation from deals that are being carried out in the market. The hotel market changes quickly, with new buyers moving in from various sectors, each with their own purchasing criteria. If the valuer is unaware of which buyers are currently active, and how they determine the price that they can pay for a property, it could adversely affect their ability to provide an accurate valuation.

There are a number of different basic methodologies that are used across the world to calculate the value of hotels, which are outlined below, and it is up to the valuer to determine the methodology that best reflects the approach taken by the market. However, all methods require, as a final step for the valuer, to stand back and review the result to ensure it is reasonable.

The profits method in more detail

There are two main methods for calculating the value of a property using the profits method: the income capitalisation method and the discounted cashflow method.

Method 1: income capitalisation method (income cap method)

One of the most common approaches adopted in calculating the value of a hotel is the income capitalisation approach, where the sustainable EBITDA is determined and a multiple is applied to these earnings to determine the value.

These calculations are all carried out in present values, and when the trading has not already stabilised and is not anticipated to stabilise for years, any growth in the income stream specifically excludes any growth attributable to inflation.

EXAMPLE: HOTEL ARABELLA

In this example, a freehold hotel with the benefit of vacant possession, it has been calculated that the sustainable EBITDA for the 100-bedroom hotel is achievable in year one as shown in Table 7.5

As can be seen, trading is anticipated to remain the same over the three-year period, resulting in a RevPAR of £60 in present values, based on a 75 per cent occupancy rate and an £80 average daily rate (ADR).

The revenue mix also remained constant, as did department costs and undistributed operating expenses.

A number of other hotels in the area of a similar quality have been sold off with a capitalisation rate of 9.0 per cent and it has been determined that this is the appropriate rate to adopt for this hotel (Table 7.6).

Table 7.5 Hotel Arabella forecast in present values

Year	*1*		*2*		*3*	
No. of rooms	100		100		100	
Rooms sold	27,375		27,375		27,375	
Rooms available	36,500		36,500		36,500	
Occupancy	75.0%		75.0%		75.0%	
ADR (£)	80.00		80.00		80.00	
RevPar (£)	60.00		60.00		60.00	
Growth (RevPAR)			0.0%		0.0%	
Revenues (£ '000s)						
Rooms	2,190.0	65.0%	2,190.0	65.0%	2,190.0	65.0%
Food	505.4	15.0%	505.4	15.0%	505.4	15.0%
Beverage	404.3	12.0%	404.3	12.0%	404.3	12.0%
Total F&B	909.7	27.0%	909.7	27.0%	909.7	27.0%
Room hire	0.0	0.0%	0.0	0.0%	0.0	0.0%
Leisure club	202.2	6.0%	202.2	6.0%	202.2	6.0%
Other income	67.4	2.0%	67.4	2.0%	67.4	2.0%
Rental	0.0	0.0%	0.0	0.0%	0.0	0.0%
Total revenue	**3,369.2**	**100.0%**	**3,369.2**	**100.0%**	**3,369.2**	**100.0%**

Year	*1*		*2*		*3*	
Departmental profit (£ '000s)						
Rooms	1,620.6	74.0%	1,620.6	74.0%	1,620.6	74.0%
F&B	363.9	40.0%	363.9	40.0%	363.9	40.0%
Room hire	0.0	100.0%	0.0	100.0%	0.0	100.0%
Leisure club	84.9	42.0%	84.9	42.0%	84.9	42.0%
Other income	23.6	35.0%	23.6	35.0%	23.6	35.0%
Total departmental profit	**2,093.0**	**62.1%**	**2,093.0**	**62.1%**	**2,093.0**	**62.1%**
Departmental costs (£ '000s)	1,276.3	37.9%	1,276.3	37.9%	1,276.3	37.9%
Undistributed operating expenses (£ '000s)						
Administrative and general	286.4	8.5%	286.4	8.5%	286.4	8.5%
Sales and marketing	134.8	4.0%	134.8	4.0%	134.8	4.0%
Property operations and maintenance	134.8	4.0%	134.8	4.0%	134.8	4.0%
Utility costs	107.8	3.2%	107.8	3.2%	107.8	3.2%
Total undistributed expenses	**663.7**		**663.7**		**663.7**	
Income before fixed charges (£ '000s)	**1,429.2**	**42.4%**	**1,429.2**	**42.4%**	**1,429.2**	**42.4%**
Fixed costs (£ '000s)						
Reserve for renewals	134.8	4.0%	134.8	4.0%	134.8	4.0%
Property taxes	158.4	4.7%	158.4	4.7%	158.4	4.7%
Insurance	33.7	1.0%	33.7	1.0%	33.7	1.0%
Management fees	101.1	3.0%	101.1	3.0%	101.1	3.0%
Other costs	0.0	0.0%	0.0	0.0%	0.0	0.0%
Total fixed costs	**427.9**	**12.7%**	**427.9**	**12.7%**	**427.9**	**12.7%**
EBITDA (£ '000s)	1,001.3	29.7%	1,001.3	29.7%	1,001.3	29.7%

Table 7.6 Hotel Arabella income capitalisation

Net base cashflow		
EBITDA in present values in year 1	£1,001,335	
Capitalised at 9.0%	11.11	£11,125,949
Gross Value		£11,125,949
Say		£11,125,000

In simplistic terms, the multiple for a property that has a capitalisation rate of 9.00 per cent (in perpetuity) is 11.11 $\frac{1}{9\%}$ or for a fixed term (say for a forty-five-year lease) it would be 10.88 $\frac{\left(1-(1+9\%)^{45}\right)}{9\%}$. In this example, there are no reductions for an income shortfall or capital expenditure.

The value is reported gross of transaction costs, in line with the guidance provided in the Red Book, and in line with local market practice.

The mathematical calculation has been rounded down to £11,125,000, which in this instance reflects market practice. Once again, whether rounding up or down, the valuer must take into account the prevalent market practice and adopt this when completing the valuation.

However, there will be occasions where the trading of the hotel is not stabilised in the first year. These reasons could include the following:

- The hotel has recently opened and is still building up its trading base.
- The property has recently undergone a comprehensive refurbishment to improve the bedroom product which will result in an increased performance.
- The proposed purchaser is planning to invest capital expenditure in the property in the future (it should be noted that a personal plan to invest in a property will only be reflected in a market valuation if the investment reflects the approach that the market would take when looking at the property) and this investment will improve the performance of the hotel.
- A change in the supply dynamics in the market has occurred, either with the closing of competitor hotels, or with the opening of new hotels in the area.
- A change in the demand for hotels in the area, either through new companies opening in the area, or changes in the road network or other transport links.
- Changes occurring to the cost structure of the hotel, for example, the need to employ additional staff, or a change in kitchen equipment leading to operational savings that need to be reflected.

EXAMPLE: HOTEL NICOLA

In this example, a very similar hotel to the one in Example 8 will generate a stabilised EBITDA of just over €1,000,000, but as a new hotel it will take three years to achieve that level (in present values) as shown in Table 7.7.

The property is also a freehold hotel with the benefit of vacant possession.

Transaction costs

There are no simple and clear written instructions regarding assessing market value, and whether such a value should be reported net or gross of transaction costs. The guidance from the International Valuation Standards states, 'Market value is the estimated exchange price of an asset without regard to the seller's costs of sale or the buyer's costs of purchase, and without adjustment for any taxes payable by either party as a direct result of the transaction', though the overriding requirement is to reflect what occurs in the market place.

As such, it is important that the valuer adopts the approach taken by the market, but also that they are extremely clear as to whether such costs have been deducted or included in the reported value.

Industry practice suggests that most investment properties – those with a lease or management contract in place – are reported net of transaction costs. These tend to be the typical transaction costs of those involved in purchasing such properties, so if a typical purchaser has the ability to pay lower transaction costs (potentially through offshore structures) then these lower costs will be deducted. In the event that the buyer has found a unique way to reduce transaction costs that may not be open to others, then typical costs tend to prevail.

Vacant possession properties are slightly different. Depending on the type of property, the size and the typical approach taken by buyers for that type of property, values can be reported either net or gross.

These projections are in present values and as such they do not include inflation.

The valuation of the property in this example shows the impact of this build-up period until stabilised trading is achieved. As can be seen, we have deducted an income shortfall of €982,800. This is the shortfall in income to the owner over the first two years (until stabilisation) against what would have been earned if the property had been earning at its optimum level in the first year.

In this example, the actual calculation is two years of stabilised income minus the actual income over the first two years: €1,001,300 + €1,001,300 – €728,000 – €291,800 = €982,800.

Method 2: DCF

The discounted cashflow method (DCF) of valuation is very similar to the income capitalisation method. In both instances, the valuer calculates the likely trading projections and applies a multiplication factor to the income stream.

There are two key differences between the methods. The DCF method calculates the trading projections in future values, so they explicitly include inflation over the period of operation.

Table 7.7 Hotel Nicola forecast in present values

Year	*1*		*2*		*3*	
No. of rooms	100		100		100	
Rooms sold	21,900		25,550		27,375	
Rooms available	36,500		36,500		36,500	
Occupancy	60.0%		70.0%		75.0%	
ADR (€)	65.00		75.00		80.00	
RevPar (€)	39.00		52.50		60.00	
Growth (RevPAR)			34.6%		14.3%	
Revenues (€ '000s)						
Rooms	1,423.5	60.0%	1,916.3	63.0%	2,190	65.0%
Food	474.5	20.0%	547.5	18.0%	572.8	15.0%
Beverage	284.7	12.0%	334.6	11.0%	336.9	12.0%
Total F&B	759.2	32.0%	882.1	29.0%	909.7	27.0%
Room hire	0.0	0.0%	0.0	0.0%	0.0	0.0%
Leisure club	142.4	6.0%	182.5	6.0%	202.2	6.0%
Other income	47.5	2.0%	60.8	2.0%	67.4	2.0%
Rental	0.0	0.0%	0.0	0.0%	0.0	0.0%
Total revenue	2,372.5	100.0%	3,041.7	100.0%	3,369.2	100.0%
Departmental profit (€ '000s)						
Rooms	925.3	65.0%	1,370.1	71.5%	1,620.6	74.0%
F&B	189.8	25.0%	308.7	35.0%	363.9	40.0%
Room hire	0.0	100.0%	0.0	100.0%	0.0	100.0%
Leisure club	49.8	35.0%	71.2	39.0%	84.9	42.0%
Other income	16.6	35.0%	21.3	35.0%	23.6	35.0%
Total departmental profit	1181.5	49.8%	1,771.3	58.2%	2,093.0	62.1%
Departmental costs	1,191.0	50.2%	1,270.4	41.8%	1,276.3	37.9%
Undistributed operating expenses (€ '000s)						
Administrative and general	225.4	9.5%	273.8	9.0%	286.4	8.5%
Sales and marketing	142.4	6.0%	152.1	5.0%	134.8	4.0%

Year	*1*		*2*		*3*	
Property operations and maintenance	94.9	4.0%	121.7	4.0%	134.8	4.0%
Utility costs	75.9	3.2%	97.3	3.2%	107.8	3.2%
Total undistributed expenses	538.6	22.7%	644.8	21.2%	663.7	
Income before fixed charges (€ '000s)	642.9	27.1%	1,126.5	37.0%	1,429.2	42.4%
Fixed costs (€ '000s)						
Reserve for renewals	94.9	4.0%	121.7	4.0%	134.8	4.0%
Property taxes	151.8	6.4%	152.1	5.0%	158.4	4.7%
Insurance	33.2	1.4%	33.5	1.1%	33.7	1.0%
Management fees	71.2	3.0%	91.3	3.0%	101.1	3.0%
Head office costs	0.0	0.0%	0.0	0.0%	0.0	0.0%
Rent	0.0	0.0%	0.0	0.0%	0.0	0.0%
Other costs	0.0	0.0%	0.0	0.0%	0.0	0.0%
Total fixed costs	351.1	14.8%	398.5	13.1%	427.9	12.7%
Incentive management fee	0.0	0.0%	0.0	0.0%	0.0	0.0%
EBITDA (€ '000s)	**291.8**	12.3%	**728.0**	23.9%	**1,001.3**	29.7%

Table 7.8 Hotel Nicola income capitalisation

Net base cashflow		
EBITDA in present values in year 1	€1,001,335	
Capitalised at 9.0%	11.11	€11,125,949
Less		
Income shortfall	€982,800	
Subtotal		€982,800
Gross Value		€10,143,149
Say		€10,150,000

As such, the multiplication factor is based on a discount rate rather than a capitalisation rate. Discount rates are also derived from market evidence, although there is typically less evidence available. Typically, a market discount rate would be similar to the product of the 'capitalisation rate' and the 'projected inflation rate'. So if the market capitalisation rate was 7.00 per cent and inflation was predicted at 2.00 per cent, the discount rate would be +/– 9.00 per cent.

The benefit of the DCF method is that explicit growth assumptions are more easily shown. In addition, most hotel investors are used to assessing value through the DCF method.

EXAMPLE: HOTEL MATTHEW

In this example, the Hotel Matthew has 170 bedrooms and offers extensive meeting facilities, including the largest ballroom in the city. It is in a well-developed and established market. Occupancy rates are high, although, given the current refurbishment programme, occupancy is quite low in the first year, at 73 per cent, rising to the typical 80 per cent once all the works have been completed (Tables 7.9a and 7.9b).

The ADR is projected to rise from CHF270.00 to CHF300.00, resulting in a stabilised RevPAR of CHF 240.00 in present values.

Rooms accounts for 66 per cent to 68 per cent, with conferencing space generating 16 per cent of turnover. The nature of the revenue mix has resulted in an EBITDA, ranging between 27.2 per cent and 32.5 per cent.

The property is valued using a 9.25 per cent discount rate, reporting a gross value of US$ 101.2 m, equating to CHF 595,294 per room (Table 7.10).

Why is a valuation required? Purposes for valuation

There are six main reasons why valuations are required detailed below.

1: *Transaction advice (both capital values and annual values)*

When undertaking to buy or sell a hotel, it is usual that both parties are keen to know the value of the property, to ensure they do not sell it too cheaply or buy it too expensively. The same is true when a lease is being agreed; if either party does not have competent advice as to the rental value of the property, then the rent may either be too high (thereby hurting the long-term profitability of the hotel operator) or too low (thereby diminishing the value of the landlord's interest).

This category of 'valuation purpose' is one of the most commonly required in the hotel world. The basis of valuation is market value, which is defined by the RICS in 'the Red Book' as follows:

> The estimated amount for which a property should exchange on the date of valuation between a willing buyer and a willing seller in an arm's length

transaction after proper marketing wherein the parties had each acted knowledgeably, prudently and without compulsion.

There are some instances when a valuer is asked to value a property subject to specific assumptions, for example, assuming that a proposed redevelopment plan has been completed, or assuming certain trading levels have been reached. These are valuations subject to special assumptions and the valuer will usually report the market value, as well as providing any market values subject to special assumptions.

It could be that these specific assumptions lead the valuer to believe that the resulting value will not reflect market value subject to special assumptions, and instead will only reflect the 'worth' to that specific owner. In this instance, the calculation must be clearly labelled as an estimate of worth and the valuer must draw this to the attention of the client.

In essence, it is a 'worth' calculation if the assumptions being requested are not those that would be reasonable for the reasonably efficient operator to adopt.

What do buyers and sellers actually want from a valuation?

As simple as it seems, what both buyers and sellers really require is information as to the market value of the property, so they know how the agreed sales price relates to the value of the property.

In most instances, the purchaser wants to be comforted that they are not paying too much for the property, and the seller wants to know that they have not agreed to sell it at too cheap a price.

2: Secured lending

In the majority of property transactions, the finance for the purchase of the property is only partly funded with equity from the purchaser; the balance is usually funded through a secured loan of some description from one of many potential banking institutions.

Most hotels are purchased with a bank providing a large proportion of the value of the property to the owner by way of secured debt, much the same way that a residential property is purchased with a mortgage. To enable the bank to lend the money on the property at competitive lending rates, they require a valuation to be carried out on the property to assess its value if it were to be sold on the open market.

The theory behind the requirement for secured lending goes straight to the heart of the theory of capitalism. To ensure that the supply of money available in the market does not keep expanding, a certain proportion of all loans have to be placed on deposit. This is required because if a bank could lend 100 per cent of the money that was placed on deposit with it, there could be a never-ending spiral in the supply of money. If the debt is unsecured, then 100 per cent of the loan needs to be placed on deposit. As the level of security of the loan increases, the amount needed to be placed on deposit decreases. A loan secured

Table 7.9a Hotel Matthew projection in future values 2016–2020

Year	*2016*		*2017*		*2018*		*2019*		*2020*	
No. of rooms	170		170		170		170		170	
Available rooms	62,050		62,050		62,050		62,050		62,050	
Rooms sold	45,297		49,640		49,640		49,640		49,640	
Occupancy	73.0%		80.0%		80.0%		80.0%		80.0%	
ADR (CHF)	270.00		297.25		315.00		322.88		330.95	
RevPAR (CHF)	197.10		237.80		252.00		258.30		264.76	
Growth (RevPAR)			20.6%		6.0%		2.5%		2.5%	
Revenues (CHF '000s)										
Rooms	12,230.1	66.0%	14,755.5	67.0%	15,636.6	68.0%	16,027.5	68.0%	16,428.2	68.0%
Total food and beverage	2,501.6	13.5%	2,863.0	13.0%	2,759.4	12.0%	2,828.4	12.0%	2,899.1	12.0%
Conference space	2,964.9	16.0%	3,523.7	16.0%	3,679.2	16.0%	3,771.2	16.0%	3,865.5	16.0%
Spa and other revenue	833.9	4.5%	880.9	4.0%	919.8	4.0%	942.8	4.0%	966.4	4.0%
Total revenue	18,530.4	100.0%	22,023.1	100.0%	22,995.0	100.0%	23,569.9	100.0%	24,159.1	100.0%
Departmental profit (CHF '000s)										
Rooms	8,805.6	72.0%	10,919.1	74.0%	11,727.5	75.0%	12,180.9	76.0%	12,485.4	76.0%
Total food and beverage	125.1	5.0%	214.7	7.5%	344.9	12.5%	424.3	15.0%	434.9	15.0%
Conference space	2,668.4	75.0%	2,748.5	78.0%	2,869.8	78.0%	2,941.5	78.0%	3,015.1	78.0%
Spa and other revenue	166.8	20.0%	176.2	20.0%	184.0	20.0%	188.6	20.0%	193.3	20.0%
Total departmental profit	11,765.9	63.5%	14,058.5	63.8%	15,126.1	65.8%	15,735.2	66.8%	16,128.6	66.8%

Departmental costs (CHF '000s)	6,764.5	36.5%	7,964.7	36.2%	7,868.9	34.2%	7,834.6	33.2%	8,030.5	33.2%
Undistributed operating expenses (CHF '000s)										
A&G	1,992.0	10.8%	2,312.4	10.5%	2,299.5	10.0%	2,357.0	10.0%	2,415.9	10.0%
S&M	1,019.2	5.5%	991.0	4.5%	977.3	4.3%	1,001.7	4.3%	1,026.8	4.3%
POMEC	1,297.1	7.0%	1,541.6	7.0%	1,609.7	7.0%	1,649.9	7.0%	1,691.1	7.0%
Total undistributed expenses	4,308.3	23.3%	4,845.1	22.0%	4,886.4	21.3%	5,008.6	21.3%	5,133.8	21.3%
Income before fixed costs (CHF '000s)	7,457.6	40.2%	9,213.4	41.8%	10,239.7	44.5%	10,726.7	45.5%	10,994.8	45.5%
Fixed costs (CHF '000s)										
Reserve for renewals	741.2	4.0%	880.9	4.0%	919.8	4.0%	942.8	4.0%	966.4	4.0%
Rates and insurance	1,667.7	9.0%	1,982.1	9.0%	2,069.6	9.0%	2,121.3	9.0%	2,174.3	9.0%
Management fees	0.0	0.0%	0.0	0.0%	0.0	0.0%	0.0	0.0%	0.0	0.0%
Total fixed costs	2,409.0	13.0%	2,863.0	13.0%	2,989.4	13.0%	3,064.1	13.0%	3,140.7	13.0%
Notional management fee – incentive	0.0	0.0%	0.0	0.0%	0.0	0.0%	0.0	0.0%	0.0	0.0%
EBITDA (CHF '000s)	**5,048.6**	**27.2%**	**6,350.4**	**28.8%**	**7,250.3**	**31.5%**	**7,662.6**	**32.5%**	**7,854.1**	**32.5%**

Table 7.9b Hotel Matthew projection in future values 2021–2025

Year	*2021*		*2022*		*2023*		*2024*		*2025*	
No. of rooms	170		170		170		170		170	
Available rooms	62,050		62,050		62,050		62,050		62,050	
Rooms sold	49,640		49,640		49,640		49,640		49,640	
Occupancy	80.0%		80.0%		80.0%		80.0%		80.0%	
ADR (CHF)	339.22		347.70		356.39		365.30		374.44	
RevPAR (CHF)	271.38		278.16		285.11		292.24		299.55	
Growth (RevPAR)	2.5%		2.5%		2.5%		2.5%		2.5%	
Revenues (CHF '000s)										
Rooms	16,838.9	68.0%	17,259.9	68.0%	17,691.4	68.0%	18,133.7	68.0%	18,587.0	68.0%
Total food and beverage	2,971.6	12.0%	3,045.9	12.0%	3,122.0	12.0%	3,200.1	12.0%	3,280.1	12.0%
Conference space	3,962.1	16.0%	4,061.1	16.0%	4,162.7	16.0%	4,266.7	16.0%	4,373.4	16.0%
Spa and other revenue	990.5	4.0%	1,015.3	4.0%	1,040.7	4.0%	1,066.7	4.0%	1,093.4	4.0%
Total revenue	24,763.1	100.0%	25,382.2	100.0%	26,016.7	100.0%	26,667.2	100.0%	27,333.8	100.0%
Departmental profit (CHF '000s)										
Rooms	12,797.6	76.0%	13,117.5	76.0%	13,445.4	76.0%	13,781.6	76.0%	14,126.1	76.0%
Total food and beverage	445.7	15.0%	456.9	15.0%	468.3	15.0%	480.0	15.0%	492.0	15.0%
Conference space	3,090.4	78.0%	3,167.7	78.0%	3,246.9	78.0%	3,328.1	78.0%	3,411.3	78.0%
Spa and other revenue	198.1	20.0%	203.1	20.0%	208.1	20.0%	213.3	20.0%	218.7	20.0%
Total departmental profit	16,531.8	66.8%	16,945.1	66.8%	17,368.8	66.8%	17,803.0	66.8%	18,248.1	66.8%

Departmental costs (CHF '000s)	8,231.3	33.2%	8,437.0	33.2%	8,648.0	33.2%	8,864.2	33.2%	9,085.8	33.2%
Undistributed operating expenses (CHF '000s)										
A&G	2,476.3	10.0%	2,538.2	10.0%	2,601.7	10.0%	2,666.7	10.0%	2,733.4	10.0%
S&M	1,052.4	4.3%	1,078.7	4.3%	1,105.7	4.3%	1,133.4	4.3%	1,161.7	4.3%
POMEC	1,733.4	7.0%	1,776.8	7.0%	1,821.2	7.0%	1,866.7	7.0%	1,913.4	7.0%
Total undistributed expenses	5,262.2	21.3%	5,393.7	21.3%	5,528.6	21.3%	5,666.8	21.3%	5,808.4	21.3%
Income before fixed costs (CHF '000s)	11,269.7	45.5%	11,551.4	45.5%	11,840.2	45.5%	12,136.2	45.5%	12,439.6	45.5%
Fixed costs (CHF '000s)										
Reserve for Renewals	990.5	4.0%	1,015.3	4.0%	1,040.7	4.0%	1,066.7	4.0%	1,093.4	4.0%
Rates and Insurance	2,228.7	9.0%	2,284.4	9.0%	2,341.5	9.0%	2,400.0	9.0%	2,460.0	9.0%
Management Fees	0.0	0.0%	0.0	0.0%	0.0	0.0%	0.0	0.0%	0.0	0.0%
Total fixed costs	3,219.2	13.0%	3,299.7	13.0%	3,382.2	13.0%	3,466.7	13.0%	3,553.4	13.0%
Notional management fee – incentive	0.0	0.0%	0.0	0.0%	0.0	0.0%	0.0	0.0%	0.0	0.0%
EBITDA (CHF '000s)	**8,050.5**	**32.5%**	**8,251.7**	**32.5%**	**8,458.0**	**32.5%**	**8,669.5**	**32.5%**	**8,886.2**	**32.5%**

Table 7.10 Discounted cashflow

Income				
2016	2017	2018	2019	2020
CHF 5,048,604	CHF 6,350,366	CHF 7,250,324	CHF 7,662,566	CHF 7,854,131
2021	2022	2023	2024	2025
CHF 8,050,484	CHF 8,251,746	CHF 8,458,040	CHF 8,669,491	CHF 8,886,228
NPV net base cashflow				
Discounted at 9.25%			101,201	
Gross value			CHF 1,200,664	

on a property may require 50 per cent of the loan to be placed, whereas a loan on a property that has an appropriate valuation showing that the lending amount is covered by the value of the asset will require a substantially lower amount of the loan to be placed in the vault. As such, a lender who has the benefit of a formal valuation will be able to offer more competitive rates to their customers because the loan is effectively less expensive to them, and the savings can be passed on to their customer.

The debt provider will usually be supplying the majority of the money for the transaction. Loan-to-value ratios for hotels differ depending upon affordability and serviceability. As such the valuer has a significant duty of care to the lending institute, and it is usual to find that the bank becomes the valuer's client, no matter who first agreed the terms for the valuation or indeed who pays the valuation fee.

What do banks need from a secured lending valuation report?

This will vary depending upon the specific terms of the loan, but in the vast majority of cases a lender will be seeking two key pieces of intelligence from the valuation report:

1. the fall-back position, in case the loan fails and the bank needs to sell the hotel;
2. the ability of the business to generate sufficient free cashflow to service and repay the loan on the agreed terms.

1: THE FALL-BACK POSITION

In the event the borrower defaults on the loan, it is essential that the bank has the capability to take control of the hotel – and they will need to know what it can be sold for. It would be unfortunate if the bank had lent US$ 50 m on a hotel, only to find it cannot be sold for more than US$ 30 m.

It is not unusual for a bank to request a number of different valuations on different assumptions, and we briefly outline the most common below.

The challenges of lending on hospitality

Tim Helliwell – Barclays Bank

I think it is interesting how the lending community's relationship with hotel valuations has evolved over the past decade. There remains a suspicion that new entrant lenders read the value in the executive summary and then file the report, safe in the knowledge that their loan-to-value assumption is fixed for the term of the loan. It usually takes an economic cycle for this lesson to be learned and contrasts with experienced lenders who are completely focused on the trading analysis, with the value itself having less relevance. As corporate diligence becomes the increasing mitigant to the relatively high leverage in the sector, investors and lenders are as equally drawn to the accountancy firms as valuers. Regulatory rules mean that valuations remain important to banks, however, the demands of the valuer are getting broader.

As the hotel sector has benefited from advisory and lending specialism for nearly a decade, there has been an interesting trend in advisory overlap – everyone thinks they can value hotels, lend on hotels, document hotel deals, etc. Valuers, and to some extent lenders, find their analysis and experience distilled into a strapline summary of '65 per cent LTV [loan to value ratio]', which in no way covers their cyclical experience (both good and bad) of supporting the sector.

Market value This is the price the hotel would sell for, as an operating business.

Market value with vacant possession Assuming the market value is subject to an operating agreement (a lease or management contract between the owner and a hotel chain) then the bank will sometimes request a valuation disregarding this operational agreement, so they are aware of the possible position should the operating agreement be terminated.

Forced sale value The majority of loan defaults occur in difficult times and lenders do not always have the luxury of time to complete a sale. The forced sale value (market value with special assumptions) generally adopts a number of assumptions (including, for example, the business being closed, no trading accounts being available, operating licences having been lost, a restricted timeframe available to complete a sale etc.) to arrive at a worst case scenario.

Market value on stabilisation New hotels take a number of years to optimise their trading potential. Lenders will sometimes request an assessment of the likely value of the hotel when trading has stabilised.

2: CASHFLOW PROJECTION FOR PROTECTION OF REPAYMENTS

Most bank loans will agree an interest payment schedule and the terms of the loan, as well as a capital repayment schedule.

It is important that the business can support these repayment requirements; otherwise, the loan may end up in default.

This is where the DCF valuation approach proves invaluable, as lenders can clearly see whether enough profit is likely to be made to support the loan repayments.

What should a secured lender look out for in a valuation report?

There are many areas where the content of a valuation report, if written well, will help a bank determine the real level of risk in lending against a hotel. A poorly written and researched valuation report will possibly not adequately highlight the inherent risks, merely state the value figure.

Below are some of the key areas where a lender should take extra care when using a valuation report for decision making.

Independence and competence Valuing hotels is by no means straightforward and a comprehensive valuation requires skill, care and knowledge of both the hotel operating market and the hotel transaction market. Banks should ensure they are using a valuer with the requisite knowledge and experience.

It is very difficult to value hotels unless you are also involved in hotel transactions. The valuer needs to be aware of who the buyers are for any such asset and what motivates these buyers, so experience of selling hotels (ideally they should work closely with hotel agents on a daily basis) is extremely valuable.

The best hotel valuers are accredited RICS valuers. The Royal Institute of Chartered Surveyors is the worldwide body that regulates chartered surveyors. The level of training required to become an RICS-accredited valuer, combined with the requirements of their continued professional development programme and regular checks on competence, ensures unparalleled expertise and competence, which is of paramount importance when providing valuation advice.

Assumptions The key to any valuation is the assumptions adopted by the valuer. Assuming that a redevelopment has been completed, for example, gives rise to a different value than if the valuer assumes the redevelopment has not been started.

As stated earlier, ideally the bank will be instructing the valuer and is therefore able to discuss their needs fully, ensuring suitable assumptions are made. Alternatively, if reviewing an existing valuation report, the bank should review the assumptions that have been adopted very carefully to ensure they are suitable for their needs.

Lack of information The bank should review the information contained in the report. Was the valuer provided with all the information they needed to carry out the valuation? Were there any omissions that the valuer has highlighted as important? Were there any omissions that the valuer has not highlighted that may have impacted on the value?

Risks attached to the deal structure, tenure and/or branding The bank should review the details of all the risks relating to the deal structure, including tenure, statutory enquiries and management. For example:

- Is the property freehold or leasehold (and if leasehold, are there any onerous terms)?
- Is the site wholly owned, or are parts of the land owned by a third party?
- Does the property have the relevant statutory consents to operate as a hotel (or are they at risk)?
- Is the management contract/lease fair? Could it be terminated by the operator leaving the owner in an undesirable position?

Risks attached to the trading environment The bank should take great care when reviewing the risks attached to the projected cashflow, as well as to the multiplication factor adopted and the resulting valuation. In particular, special care should be taken concerning the following:

- How does the projected future trading compare with the available historic trading data, and why are there any differences? Have they been adequately explained?
- Are there any proposed changes to the hotel supply in the area, and have these been adequately reflected in the projected trading?
- Are there any likely changes to local demand for hotel accommodation, and have these been adequately reflected in the projections?
- Are the adopted assumptions reasonable?
- How has the multiplication factor (capitalisation rate or discount rate) been calculated, and does it reflect current market levels?
- How does the capital value compare with comparable transactions?
- Are there any risks with the project/location/market sector that might affect capital values, over and above the trading risk inherent in the project?

All of these factors should be discussed and any issues should be clearly identified in the valuation report.

3: Property taxation

In a number of countries, property tax is based upon the notional capital or rental value of the property and as such, both the government and the owner need to be able to calculate the liability accurately to ensure that neither too much nor too little tax (from their respective positions) is being collected.

4: Company accounts

A number of listed companies choose to revalue their hotel assets on a regular basis and these values are reported within their company accounts. Without accurate assessments of the value of the properties it is quite likely that the share price of the company could not be able to be accurately assessed.

The 'Red Book' provides a good basis for how to carry out a valuation of a hotel (as a non-specialised property) for inclusion within reported company accounts. Valuations for inclusion in financial statements prepared in accordance with 'generally accepted accounting principles' (GAAP) shall be on the basis of either:

Market value with existing use (EUV) for properties that are owner-occupied for the purposes of the entity's business; or

Market value for property that is either surplus to an entity's requirements or held as an investment. The definition of market value with existing use is:

> The estimated amount for which a property should exchange on the date of valuation between a willing buyer and a willing seller in an arm's length transaction, after proper marketing wherein the parties had acted knowledgeably, prudently and without compulsion, assuming that the buyer is granted vacant possession of all parts of the property required by the business and disregarding potential alternative uses and any other characteristics of the property that would cause its market value to differ from that needed to replace the remaining service potential at least cost.

What do shareholders want from a valuation?

To be able to ascertain the asset value of the company underlying their investment.

5: Internal purposes

It is sometimes important for companies to review their assets for purely strategic reasons; in these cases, the valuer will be looking to meet the requirements of the client and will not be bound by the usual valuation guidelines, unless of course the client requires a market value.

There are many times when a client requires a 'valuation' to be carried out that is specific to their requirements or looks to include very specific assumptions that would not be typical of the market in general. In this instance, the valuer will not actually be carrying out a valuation but will be undertaking an 'estimate of worth'. There is no problem with carrying out any number of worth calculations, and the client can request, and the valuer can carry out, the most unlikely calculations as long as the valuer is very clear throughout the report and when discussing the figures with the client that it is a calculation of worth and not a valuation.

Worth is the value of property to a particular owner, investor or a class of investors for identified investment objectives. This subjective concept relates to a specific property, to a specified investor or group of investors and/or an entity with identifiable investment objectives and/or criteria.

The valuer may also include details of the market value of the property so that the client could see the relationship between the market value and the estimate of worth that has been carried out.

What do hotel companies/investors want from their worth calculations?

The requirements for their calculations are likely to be as varied as the special assumptions that are requested, but normally the company/investor is keen to know the potential impact on the asset valuation of certain courses of action.

6: *Company taxation*

It is necessary to determine the depreciation required on the asset for accounting purposes, to ensure that the correct element is deducted for tax purposes. A company will be able to write-off an element of the value of its assets over their expected lifecycle to ensure that it is able to calculate its company tax liability appropriately.

This is a specialist area and can vary from one tax jurisdiction to another; as such, it is essential that the valuer is aware of the exact taxation guidelines that are in place in the location that they are working.

8 Due diligence

The importance of due diligence

Due diligence is a technical term for the research done as part of a property transaction, usually carried out by a team of experts. It covers many disciplines including legal issues (including statutory enquiries), financial analysis and review, operational assessment, construction and condition surveys and of course valuation.

It doesn't matter how many times a purchaser visits the property to review what is happening on the ground or what is proposed, it is still essential that expert due diligence is commissioned. As this book is mainly about assessing the value of a resort or hotel valuation, due diligence is obviously very important, whether to raise finance for the purchase or to provide peace of mind that the price is not too high.This advice should always be provided by a RICS-qualified valuer with the relevant market experience.

However, there are many other types of expert advice that the purchaser may need to take as discussed below.

Structural surveys

It may be that the property is old, in a state of potential disrepair or in need of modernisation or refreshment. Sometimes it will be useful to have a full structural survey completed so that future plans for the property can be mapped out properly, with an accurate estimate of what cost might be incurred.

Financial

It is important to commission financial due diligence on the property. A valuer will look through the accounts to assess the value of the property, but this will not be a detailed analysis and cross-checking of the actual accounts provided. Financial due diligence will provide some comfort that the stated accounts are accurate, that forward bookings are accurate, deposits are still held in place, that no tax liabilities exist (assuming a company purchase is being considered rather than a straightforward property transaction) and that everything is as the

USALI

The hotel industry is unique in that it has a single worldwide accounting standard that is used for most hospitality businesses in most parts of the world. Known as 'the Uniform System', or to use its full name, the 'Uniform System of Accounts for the Lodging Industry', USALI has been used since 1926 and provides a standard basis for all hospitality products to generate a standard form of accounts.

The latest edition, the eleventh, came into reporting force in January 2015. It has been revised from the previous edition to reflect changes in industry practice and address issues that arise as the industry develops. The main changes, excluding renaming certain headings, are adding IT systems as a fifth undistributed expense department, and adding a 'cluster services' expense category.

A large number of hotels and hotel companies are still reporting based on the tenth edition and, when benchmarking a property's performance in comparison to its peer group, it is important to ensure that the comparison is based on other properties using the same reporting standard.

purchaser has been led to believe by the vendors. The financial advisor should also be in a position to start your tax planning straightaway to ensure the correct structures are put in place to suit the purchaser's circumstances.

Legal

As with any property transaction, it is essential to carry out legal due diligence. This can be confined to straightforward conveyance advice (is it freehold with no rights of way, good marketable title, etc.), or it can extend further into all the contracts involved with the purchase (staff employment contracts, future booking contracts, etc.).

The levels for such due diligence will depend upon the type of the property that is being purchased. This section covers legal and condition due diligence, with the following chapters exploring other types of due diligence.

Legal due diligence

Legal due diligence covers a wide variety of areas in a hotel transaction. The legal team provides expert advice relating to various matters, outlining what is involved in the transaction, including what the liabilities going forward will be, and usually providing commentary on any matters that are slightly unusual in the transaction. The key areas where the legal team will advise are usually as follows:

- title
- management contracts or operating leases
- employment
- operating licences
- planning and consent for commercial use.

The quality of the title

The value of the hotel will be affected by its tenure as the right to hold a property indefinitely (freehold title) is more valuable than only being able to hold the property for another twenty years before it reverts back to the landlord, with or without the benefit of any statutory protection (e.g. Part II of the Landlord and Tenant Act 1954 in the UK).

If the property is leasehold, there could be a number of significant clauses in the lease that could affect value and it will be essential for the investor to review either the lease or a report on the title produced by a reputable lawyer. Anything contained in the lease that will deter someone from wanting to own the property could lead to a lowering in the value of the property, as the competition for the property is likely to be more restrained. However, there are a number of relatively common terms that have an impact on value.

These key terms will include the following.

The term of the lease

In most instances (if the other lease terms are not onerous), the longer the term, the more valuable the property interest is. The value attributable to a property is generally enhanced by the longer that property can be 'enjoyed', so typically a lease where the property can be used for forty years is more valuable to the owner than a lease where the property can only be used for five years. However, if the obligations of the lease are onerous, for example, the rent is too high to be affordable, then the shorter the term the better from the investor's perspective.

The demise of the property

What is included in the lease in terms of land and buildings will directly impact on the value of the property.

The rent payable and any rent review provisions

The rent payable will impact on the value of a hotel, as it generally impacts on the earnings the investor can expect from the hotel operation. Typically, a higher rent will mean a lower value (to the tenant) because any rent that needs to be paid to the landlord will be deducted under fixed costs in the profit and loss account (P&L) and the EBITDA will be reduced by this amount.The timing of rent reviews and how the rent is assessed will also impact on the value of the property. The more favourable the rent review provisions for the landlord, the less valuable the tenant's interest.

The user clause

It is common for leases to specify a use for a property and to forbid certain uses, and this can have an impact on the value of an asset. For example, the sale of alcohol or the use of the property for holding auctions are commonly excluded, both of which could impact value if they were areas of trade that the hotel wished to become involved in.

The repairing liability

It is essential for the hotel to undertake necessary repairs enabling it to attract customers. A hotel has a higher level of care than for most other property classes because hotels effectively need to attract new tenants every night. Other property classes only need to be in good condition when they are being let. An overly onerous repairing clause can however adversely impact on the value of a property. Some leases specify a certain proportion of turnover to be spent each year on the maintenance of the property, and if this percentage is too high, it can detract from the value of the property.

The insuring liability

Any unduly onerous provision could adversely affect the value of the property.

The provision for alterations

When a landlord has a right to prohibit alterations, whether absolutely or subject to certain provisions, this can have an impact upon the value of a property. Hotels are constantly evolving and unless they can adapt to the requirements of their clientele their trading will suffer, which could affect the value of the property.

The alienation provisions of the lease

The alienation provisions affect the transfer of part or all of the property (the lease interest) to different tenants, sub-tenants or under tenants. They include the right to sublet part or all of the property, as well as the right to assign part or the whole of the property. Simplistically, assigning is transferring your entire interest in the property (your lease) to another party, while subletting is creating another, lower, interest in the property which is passed to the third party, while you still retain a legal interest in the property. The more restrictive the alienation clause, the smaller the potential market for the property, and therefore in certain circumstances this will result in a lower value. In the event of a blanket restriction on assignment of the whole of the property, effectively the property cannot be sold and therefore there would be no 'market value' to that asset, if such a restriction were legally enforceable.

Forfeiture provisions

Where the forfeiture provisions are unduly onerous or impact on the ability of the leaseholder to secure funding for the property, it will have an adverse impact on the number of potential buyers for the property, and ultimately on the value.

There will be other key considerations contained within the clauses of the lease depending on where the property is located and the legal system that regulates such leases.

However, it is not just as simple as the difference between leasehold and freehold. The quality of the title will also be important in determining the value of the asset. In this respect, reports on title are invaluable, as they have been prepared by experts and can be relied upon by the investor. For example, it may be difficult to determine whether the boundary of the land on the title documents extends all the way to the public highway. Should the boundary not abut the public highway, it may be that access to the hotel is under the control of a third party (sometimes known as a ransom strip). Your lawyer should be able to advise on whether the hotel property connects with the adopted highways and services. The impact of access issues to a hotel or resort can be substantial in certain circumstances. The investor will need to consider their ability to rectify any title defect, to secure another access (and the cost of doing so) or alternatively the cost of an insurance indemnity policy when looking at the impact such a defect could have on the value of the property.

Another typically difficult default in the title of a property may be in the provision of a restrictive covenant. Restrictive covenants are negative covenants and in English law can be quite difficult to expunge. Restrictive covenants that are quite common in hotel documentation include the prohibition of specific uses (including forbidding the sale of alcohol, casino use, residential accommodation, car parking, auctions and even for the use as a hotel), all of which could have a detrimental impact on the trading potential of the hotel and therefore its value. It could be the case that the hotel has a more valuable alternative use and so any prohibition on the land or buildings use for 'other uses' would also detract from its value in that instance.

Other areas of defective title include where the quality of the title is not good and insurance needs to be taken into account. For example, in large areas of Eastern Europe, there are problems with the title of property as the past owners of the land were forcibly evicted from the land either during the Second World War or during the following communist regimes. The claims for repatriation from two potentially 'displaced owners' of these lands has led to uncertainty in the quality of the legal title available for prospective purchasers.

Alternatively, only part of the land may have good clean title, with part of the property uncertain of its title position. In that case, the investor will need to 'take a view' on the importance of that part of the site to the overall operation.

If it forms a fundamental part of the hotel, the investor will need to review the possibility of rectifying the title defect, taking out indemnity insurance or replacing the affected part of the property on another part of the site. If it is part of the grounds, the impact of the 'poor' title may not be so significant.

Management contracts or operating leases

The lawyer will outline all the key terms of any operating agreements (whether leases or management contracts), preferably advising the investor when such clauses are not standard. Non-standard clauses may have an impact on the level of interest from potential investors and therefore can impact on value. They will also outline the key responsibilities and implications of the terms contained in the agreement. A management contract is an unusual structure for many types of investors as it transfers most of the 'operational risk' across to the investor, at the same time as transferring the property risks as well. The investor behind a management contract will be employing the operator to manage the property on their behalf. This means that the staff are all employees of the investor (along with all the relevant employment regulations and legislative issues this raises). In addition, the repairing liability for the property is rested squarely on the shoulders of the investor, and if the manager (or management company) is not very good then the investor's returns will be adversely affected.

A management contract should always be viewed as a partnership between the owner and the operator rather than in any other way. If one side has too much control of the partnership (the terms are too favourable to one party) then it tends to fail as a relationship. The investor needs to earn a reasonable return on their investment, while operators need to generate reasonable management fees. If either of these 'returns' is too low, one party will be dissatisfied with the agreement and will have no reason to wish to continue it. A management contract is similar to a turnover lease, but has the advantage of avoiding a balance-sheet liability for the operator, while providing easier provisions for re-entry for an owner when the operator is not performing. However, there are three key investment differences between a turnover lease and a management contract:

1 employment liability risks
2 repairing liability risks
3 additional operational risks (cost risks).

Some of the key terms in a management contract are as follows:

The 'term', or how long the agreement will last

Most terms tend to last between ten and twenty-five years. Any period outside this range could have an impact upon the value of the investment. If the management contract 'adds value' to the property, then the longer the term the better. If the property is highly desirable and likely to attract multiple potential operators, then a shorter term may add value. Whether any extensions to the term are mutual or at the discretion of the operator can also have an impact on the value of the investment.

The level and detail of the management fees

The cost of the management of the hotel by the operator will impact upon the cashflow generated by the owner, and as such will have an impact upon the value of the investment. It is usual to see a base fee (which is a percentage of turnover) and an incentive fee (usually a percentage of AGOP). The level of fees will vary depending upon the locality but fees in the region of 2 per cent to 3 per cent of turnover as a base fee and 8 per cent to 10 per cent of AGOP as an incentive fee are typical. AGOP can be calculated in many ways, depending on the contract. It is advantageous for the investor to deduct as many 'fixed costs' as possible before calculating the incentive fee, if that can be agreed.

Certain contracts allow for incentive fees to be subordinated to debt service or some other agreed performance criteria. Subordinated fees may be waived or accrued, depending on the relative strength of the negotiating parties when the agreement was being drawn up. In certain circumstances unpaid fees become debts, chargeable upon the property which needless to say has an impact upon the value of the investment.

The level of the fees, along with any subordination of fees, will have an impact on the desirability, and therefore value, of the property.

The operator system fees

The way that 'other charges' are handled will affect the income generated by the investment, and therefore its value. Reservation fees can be charged per booking or per room-night, or they can be a percentage of revenue generated or a fixed price per booking. They can be charged at the time of booking or when the guest arrives, sometimes reservation fees accrue in the event of no-shows. All of these factors can have an impact on the value of the interest to the investor.

Shared services

It is usual for the owner to have approval of those services which are to be shared with other hotels operated by the operator. These services should be detailed in the agreement. If there is no control over such shared facilities (for example, the auditor for a competitive hotel run by the same management company has the right to review the hotel accounts), it can have a detrimental impact upon the value of the investment.

Performance test criteria

Many management contracts contain no performance test criteria at all, although the better-drafted contracts (from an investor's perspective) provide the right for the owner to terminate the contract, without compensation, should the operator consistently fall short of the performance test.

A performance test will usually specify a target that the management company would have to attain if it were not to fall foul of such a provision. For example, the management company may be tasked with achieving 90 per cent of the agreed budget (possibly in terms of revenue and GOP), or achieving 90 per cent of the RevPAR of an agreed competitive set of comparable properties. It is usual that the performance test would require repeated failure, rather than a one-off 'underperformance', and typically this is for three consecutive periods. Choosing the competitive set for benchmarking purposes is quite subjective and these can sometimes be unsuitable. Any defect in such a competitive set can adversely affect the value of this clause to the investor.

Owner's right to sell the property

The owner should normally be free to sell the hotel without the operator's consent, although many older agreements have a blanket restriction on the owner selling the hotel without the operator's approval. If a purchaser wishes to change management companies, or is itself a management company, there is usually a right to terminate the management agreement, with compensation paid to the operating company. In some management contracts, the operator has a right of pre-emption and can take over the deal after it has been negotiated. This can have a dramatic impact on the value of the investment as it can substantially lower the amount of investors prepared to go through the due diligence process. In this instance, such investors know that at the end of the process it may have all been in vain, despite the truest intentions of the vendor, and they could be left with the very real expenses they have incurred through the process.

Owner to approve budgets

Most management contracts state that the operator will provide the owner with detailed operating budgets, a breakdown of proposed capital expenditure and a marketing plan for the owners' approval, each year.If the owner does not have this right, it could adversely affect the owner's investment.

Owner to have the right to approve the general manager and financial controller

Most contracts will allow the owner the right to approve the appointment of the property's general manager and sometimes other key personnel like the financial controller. Without the ability to question the appointment of the general manager, the owner is in a weaker position to ensure the optimum performance of the investment.

Redevelopment in the event the property is destroyed

Some operators request a clause that obliges the investor to rebuild the property in the event it is destroyed. The owner should have the freedom to choose not to rebuild the hotel in this instance, as having the freedom can sometimes lead to a potentially higher alternative use and therefore value.

Restrictions on competition

The contract may limit the number of hotels that the management company can operate within a 'competitive' geographic locality to the subject hotel. It may be that the restriction is merely on the use of the brand name of the unit. The greater the restrictions, the more control the investor has over the future openings operated by their partner affecting their investment. The logic is that the investor has agreed to have a 'Holiday Inn' brand which they feel is beneficial to the trading potential of their hotel. If another 'Holiday Inn' opens very close to the property then some, or all, of that benefit is lost, and the reason for entering into the agreement on those terms has been lost. Alternatively, or in addition, the owner may have a priority listing in the operator's reservation system (i.e. always first choice) or the right to approve any additional use of the operator's name.

Bank accounts to be the property of the owner

The bank accounts are usually held in the name of the owner and not 'held in trust' by the operator. The owner will usually also have the right to approve the signatories to the account. If they do not have these rights, it can be detrimental to the value of the investment.

Branded operator's equipment

This should be kept to a minimum to avoid any additional expense on behalf of the investor on termination of the management contract. If the level of branded equipment is unusually high, then it could have a detrimental effect on the residual value of the investment.

Arbitration

Most contracts allow for arbitration in the event of disputes between the parties. However, where the chosen seat of such arbitration is, and the applicable law, can have quite an impact on the value of the investment.

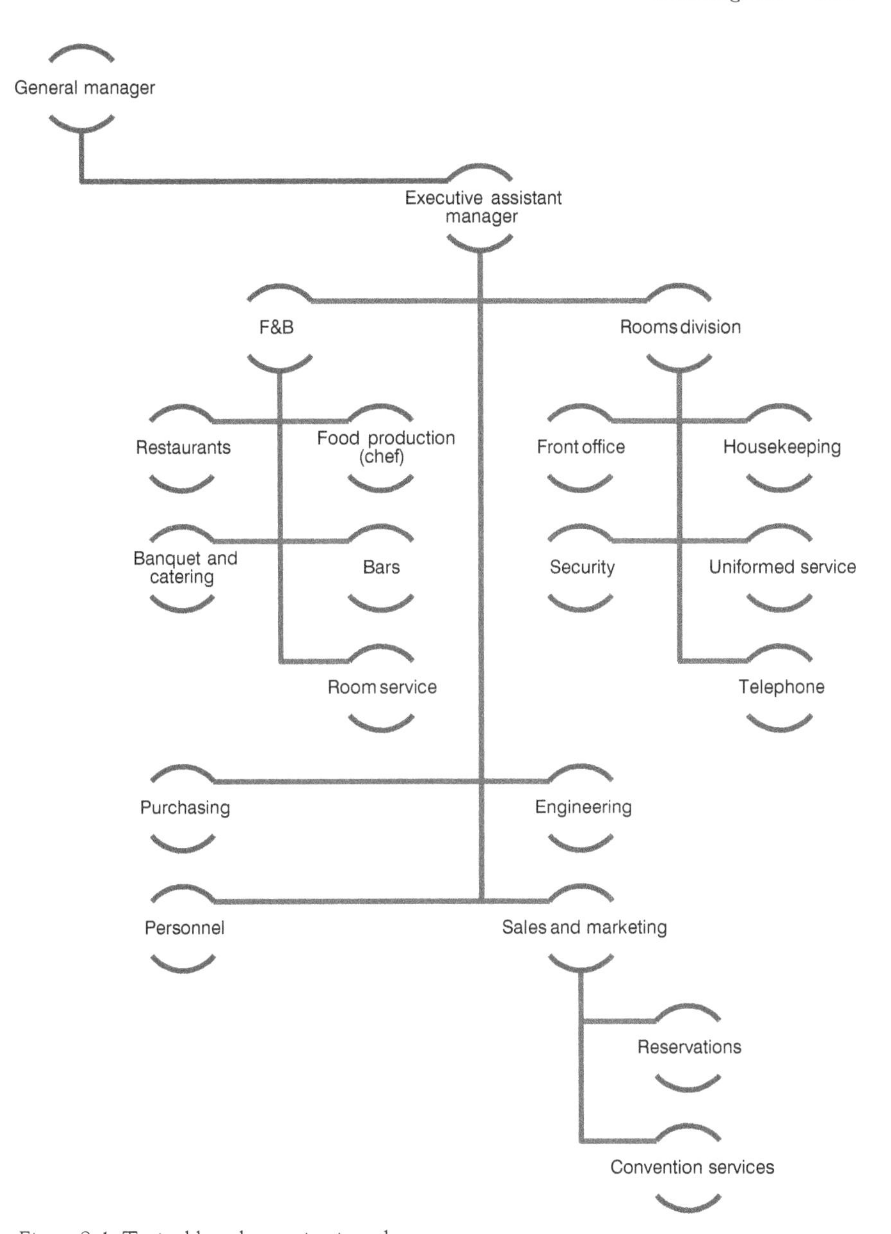

Figure 8.1 Typical hotel organisation chart

Employment

Hotels and resorts need staff to operate. Without them, the property would be unable to produce any income for the investor, which would have an adverse impact on value. As such, staff tend to be transferred along with the property in a hotel or resort sale.The lawyer will review all the employment contracts of the transferring staff to ensure all they are standard. If staff contracts are unduly onerous (either in favour of the staff or at the expense of the staff), it can have an impact on the value of the asset. Also, if the hotel employs too many staff, it can have an impact on profitability, and therefore value. The level of staffing will vary depending on the type, quality and the location of the hotel. Expert advice is usually required to assess whether the hotel has too many staff.

Statutory enquiries

The legal team will usually also review the statutory obligations ensuring all the required consents are in place, and that the property is not in danger of closing down as a result of being in breach of statutory requirements.The statutory requirements that need attention are quite diverse, and vary from country to country, region to region. Areas reviewed usually include the following.

Property taxes

Property taxes will require research (to be able to assess the level of fixed costs of the property). These vary significantly property to property and can have a major impact on the profitability of the hotel. Most countries throughout the world have different methods of assessing the level of taxation, all based on very specific principles. In the UK, the rating liability changes each year, with a revaluation of the property every fifth year and the rates payable as a proportion of this rateable value being assessed on an annual basis. However, in the Ukraine, there are no local property taxes and in Russia taxes are based upon a notional ground rent charge. It is very important that the investor pays attention to the specific details of the relevant property tax liabilities, as these can, in some circumstances, change quite radically from historic levels, impacting significantly on the value of a hotel or resort.

Planning and highways

The planning process varies across the world but most systems are based upon similar principles, where land is zoned for specific uses and certain buildings and areas are more heavily protected than others in an attempt to preserve buildings of particular historical importance, or areas of particular character. It is important that the investor knows the zoning of the land in which the property is built (along with any density restrictions where these apply) and the degree to which the property is 'protected' or 'listed'. These factors could impact upon the potential redevelopment/extension of the property in the future, or indeed impact on the required repairs and maintenance expenditure for the hotel.

The investor and their professional team will need to look into the planning history of the subject property to ensure that it has consent for its current use and consent for all the building structures on the land, and whether any outstanding planning conditions remain. Although there are sometimes statutory time limits that apply for presumptive changes of use and for retrospective planning consent, it would be usual to see a hotel that was either trading outside its planning use, or did not have consent for the structure as being worth less as a result of these defects. A review of the planning history of the property will enable the investor to determine if there are any outstanding planning decisions pending, or indeed valid planning consents that have not been undertaken. It can also be helpful if the investor is told about the development potential of the hotel, to review the relationship with the Local Planning Authority. For example, if the hotel has previously tried to extend the bedroom accommodation only to be refused on specific grounds and if these grounds still apply, it may be difficult to realise such development potential.

It will also be important to review the planning position of sensitive sites near the property. If, for example, the vacant lot next door has just been granted consent to build an incineration unit or an abattoir, that could have an adverse impact on the future trading potential of the hotel, and therefore its value.

The investor also needs to consider major planning changes in the area, for example, the provision of a new conference centre (which could increase business) or the provision of a new bypass (which could lower the visibility of the hotel and therefore the volume of its business), and take any changes into account when looking at the potential business and value of the property.

Operating licences

Each country then has a number of operating licences or certificates that are required if the property is to be operated as a hotel. These may include fire certification, opening certificates, licences to sell alcohol, licences to extend usual opening hours in the restaurant or conference rooms, licences to carry out weddings and civil ceremonies, restaurant licences, etc. As part of the due diligence process, the legal team and the valuer will be expected to determine whether these are all in place.

Health and safety

It is usual for most geographic jurisdictions to have a health and safety audit of hotels, whether checking the quality and cleanliness of the kitchens, or if the property is 'safe' in the event of a fire, or whether they are likely to expose their guests (or staff) to Legionnaires' disease, for example. Ideally, the investor and their professional team should make contact with such organisations to see if the hotel has complied with all of its requirements or is about to be closed down through non-compliance.

Setting up a hotel in Singapore

Every country has different requirements which a hotel owner/developer is obliged to meet, if they are to be permitted to operate a hotel. Legislation varies dramatically across the world, so it is essential to look into the requirements of the specific jurisdiction. Below is one example, a hotel in Singapore, just for illustrative purposes, to demonstrate the many stages sometimes required before a hotel can become operational. It should be noted that buying an operational hotel does not guarantee all necessary licences are in place, and it is essential that the legality of the operation is reviewed before such a purchase is made.

Certificate of registration and hotel keeper's licence

This licence is essential for anyone wishing to operate a hotel. Certain conditions needs to be fulfilled before such a licence can be obtained. These include:

- business incorporation (registered in Singapore);
- approval of premises (authority for the premises to be used as a hotel);
- appointment of a hotel keeper (a CEO or GM must be appointed to operate the hotel);
- advertising the application (the application must be advertised locally so objections can be registered);
- environmental clearance (to show the use will not be harmful to the environment);
- fire safety approval (a fire certificate is required);
- approval from the Building Construction Authority (a temporary occupation certificate or Central Statutory Completion certificate is required.

If a licence is granted, it needs to be renewed annually.

Non-residential TV licence

There is a requirement for this licence if any device, TV, personal computers or tablets are located on site and have the ability to receive broadcasting signals.

Public entertainments licence

This is required when any events are held where public access is available, whether the event is free or a charge is made for entry. This might include any type of show, karaoke, competitions or quizzes.

Arts entertainment licence

An arts entertainment licence is very similar to the public entertainments licence, but tends to concentrate or more artistic events such as plays, poetry readings and art or photographic exhibits.

Public establishment licence

Such a licence is required to allow the hotel to provide F&B facilities.

Copyright permit

This certificate will be required if the hotel intends to play music or videos anywhere on the premises.

Halal certification

If the hotel intends to provide halal food, a licence from the Islamic Religious Council of Singapore is required.

Massage establishment licence

If the hotel wishes to offer any spa services, including massage, reflexology or nail treatments, then a massage establishment licence is required.

Money changer licence

A licence is required to allow the hotel to offer money-changing facilities.

Permit to buy, store and use hazardous substances

A certificate is required to store many common cleaning products, as they are deemed hazardous when stored in commercial quantities.

Swimming pool licence

A separate licence is required for a swimming pool.

Tobacco retail licence

If vending machines are installed, or there is a shop that sells tobacco, then a specialist licence is required.

Registration to import food and food appliances

Certain food products and appliances require registration with the Agri-Food and Veterinary Authority of Singapore.

Petroleum/flammable materials storage licence

A one- or two-year licence can be obtained if storage facilities are deemed safe.

Advertising licence

Sky signs and electronic billboards require a further licence.

General radio communication licence

This licence is required for receiving or transmitting messages by radio communication.

Localised private network licence

The use of walkie-talkie services within a localised area would require this licence.

Localised radio communication licence

Pagers, walkie-talkies and alarms would trigger a requirement for this licence.

Wide area private network licence

The use of walkie-talkies or other such devices in a wider area (non-localised) would require this licence.

Licence to discharge trade effluents

Discharging trade effluents into a water course or drain requires this licence.

Approval for premises

Any changes to the design or alterations to the structure of the building require approval from the building authority.

Health and safety is of paramount importance to many investors and operators, and the value of such due diligence cannot be overstated. Therefore, it is vital that the investor has an expert team that is well-versed in the requirements of the specific jurisdiction in which they are investing.

Environmental surveys

An environmental survey is required to ensure that there are no environmental factors that will impact upon the operation of the property, no contamination

Table 8.1 Example of a contaminated land risk assessment

Item	*Assessment*	*Risk category*
Potential for statutory liability and designation as contaminated land	Potential pollutants have been identified. However, there is no evidence to suggest that any significant environmental concerns exist.	Low
Potential for third-party liability	There is the potential for a degree of residual contamination on the site from historical uses. However, this is unlikely to be significant given it was redeveloped in the 1980s and currently comprises building cover and hardstanding which should restrict the infiltration of rainwater and leaching of mobile contaminants, if present, off site.	Low
Risk of contaminated land liability for owner	There is the potential for a degree of residual contamination of the site on the basis of former land uses, though given the lack of regulatory concern and continued commercial use, the risk of contaminated land is considered reduced.	Low

that will impact it or surrounding properties, or any other factors that could impact on value. In many legal jurisdictions, the owner of the land is liable for any future contamination that may occur from that land, even when the initial contamination occurred when that party did not own that land. As liabilities for contamination can be extremely high, most purchasers undertake some form of environmental survey. Surveys typically range from the following:

- desktop analysis
- desk-based analysis combined with site survey
- ground testing, whether random samples or more comprehensively concentrating on potential trouble-spots.

Where there are outward indications that contamination may be present, whether because of the historic use of the site or because of physical evidence, it is more likely that ground, soil and water tests will be undertaken.

A typical survey will discuss the likelihood of contamination, the likely level of such contamination, recommended remedial works and the potential cost of such works.

Needless to say, any risk (and cost) implications from the environmental survey could have an impact on the value of the property. In extreme situations, the cost of remedial works to the contamination can be greater than the value of the uncontaminated asset, leading to a nil or negative value.

Table 8.2 Regulatory data

Item	*0–249m*	*250–500m*	*Comments*
Contaminated land register and notices	0	0	
Registered landfills	0	0	
Closed landfills	0	0	
Registered transfer/treatment centres	0	0	
Closed transfer/treatment centres	0	0	
Authorised industrial processes	1	1	LAPPC permits are recorded for dry cleaning processes located 190m to the west and 320m to the north east of the site.
Fuel stations	0	0	
Licenced radioactive substances	0	0	
Enforcement prohibitions or notices	0	0	
Discharge consents	0	0	
Pollution incidents	2	2	The closest incident occurred 175m to the east of the site in December 2005, comprising the release of oil into the drains.
Consents issued under the Hazardous Substances Act	0	0	
Control of major accident hazards	0	0	

Condition surveys

Typically, before purchase a buyer will commission a condition survey to ensure that the property is in the expected condition and that any required capital expenditure has been built into the purchase price.

The detail required in the building survey will depend on the age, design and to some extent the external condition of the property. However, the value (or purchase price) will also be a factor, as it is generally considered prudent to commission a full survey of the property if the overall investment is significant, even if there is no outward sign that defects are present. The scope of a survey varies and is determined by the parties involved, usually the potential purchaser and their lenders. Surveys are usually arranged in three key components:

1 building fabric
2 mechanical and electrical services
3 building finishes.

Figure 8.2 Condition due diligence

Figure 8.3 Environment due diligence

Table 8.3 Major remedial items

Item	Time scale Yr 1 £	Yr 2–5 £	Yr 6–10 £	Comments
Structure and fabric				
Barrel vault roof		75,000		Repairs or replacement
Low-level flat roof	25,000	35,000		Showing signs of deterioration
High-level flat roof		35,000		Undertake repairs
Car park	50,000	25,000	25,000	Concrete repairs to car park
Concrete cladding panels		22,500		Replace worn sealant joints
External staircase	5,000			Monitor and repair cracked brickwork
Internal finishes	8,750			Repairs to damp affected walls
Subtotal	88,750	192,500	25,000	
M&E services				
Electrical distribution	50,000			Replace obsolete electrical distribution boards
Rooftop extract		15,000		Replace rooftop extract fans
Improve car park lighting	35,000			Install additional lighting
Replace controls		25,000		Upgrade plant controls
AHU replacements		200,000		Commence change-out programme
Kitchen service lift	55,000			Replace car and motor
Mains water pumps		20,000		Replace aged pumps
Subtotal	140,000	260,000	0	
FF&E				
Upgrade bedrooms	125,000	500,000	500,000	
Change carpets and lighting in public areas		125,000		
Refurbish bar area	45,000			
New carpet in restaurant		40,000		
New lighting in meeting rooms	15,000			
New sound system in ballroom	25,000			
New equipment in gym		30,000		
Subtotal	210,000	695,000	500,000	
Total	438,750	1,147,500	525,000	

The findings from a condition survey will vary in presentation, and indeed in the scope and detail provided, but an example of a typical report might be something like that outlined in the example in Table 8.3.

Any unexpected capital expenditure requirements will have an impact on value, and may be used to renegotiate the agreed purchase price.

Due diligence

Paul McCartney – DAC Beachcroft

When contemplating an acquisition or investment, the buyer should always bear in mind that under English law, the basic principle is *caveat emptor*, or, 'let the buyer beware'! While the buyer may seek to protect themselves in the acquisition agreement by seeking warranties from the seller about the business or assets in question, in practice these may offer limited protection. Therefore, it is of critical importance that the buyer conducts as thorough a due diligence exercise as is possible under the circumstances in order to gain an understanding of the strengths and weaknesses of the business being acquired. The results of the exercise can assist the buyer in working out what risks there are with the purchase, allowing for negotiation on how those risks may be allocated, what contractual protections are required, what price adjustments should be made and, indeed, whether the proposed deal is worth the risk at all.

Problems can come in many forms: a black hole in the accounts; inter-company debt inadvertently left in the corporate target; impending litigation; impending loss of a major customer or supplier; skimping on property maintenance, particularly major capital expenditure (capex) items; the fact that the hotel is built on land subject to restrictive covenants against hotel use; or even (on one occasion) that the main front entrance to the property is situated directly on top of a former mineshaft! Let the buyer beware!

9 Financial due diligence

Understanding the business

Financial due diligence is essential to check that the information being provided to the buyer is accurate. As such, the very first step is to check that the accounts are accurate. This is usually carried out by auditors, and it is important that good quality auditors, familiar with hotels and resorts of the relevant type, are used for this function. The process will involve checking receipts, deposits, tax statements, bills of sale and indeed going through the whole 'paper trail'.

Once this has been completed, the second step, which is of greater interest to the investor, commences. The analysis of the accounts will lead the investor to understand in great detail how the property trades, and where opportunities for the future may lie.

Analysis of the accounts

The analysis of the historic accounts is an exceptionally good way to start to understand the property and the peculiarities of how it has historically traded. However, the investor must keep at the forefront of their mind that the historic accounts only show how that particular operator has been running the hotel, and that may not reflect the position of how the new owner would operate the property.

A basic understanding of company accounts is required and it is not the intention of this book to go into the specifics of book-keeping. However, it is important to draw your attention to the Uniform System of Accounts for the Lodging Industry (USALI) which has been used for hotel accounting since 1926 and is still the industry standard today.

In a nutshell, USALI determines where specific revenue and costs should be accounted for in the profit and loss accounts, so that all hotels have a relatively comparable set of accounts for benchmarking.

For example, if one hotel accounts for its room profits after having deducted sales costs (which are usually attributed to undistributed operating expenses under sales and marketing) then its profit margins will be lower than would be anticipated when looking at other similar hotels, which could confuse the investor when looking into the departmental profit margins.

Typical layout of a hotel's profit and loss account

Table 9.1 shows a typical summary sheet, which outlines the main headlines of the accounts. This summary would then usually be backed up by all the detailed accounts for the hotel.

For those inexperienced with hotels, these accounts can look rather daunting. However, if taken one section at a time, they are quite clear and relatively easy to understand.

The first part of the accounts details a statistical analysis of the performance of the hotel, concentrating on what is traditionally the most important single revenue stream into the hotel, the rooms' department. It outlines the occupancy level, the average daily rate (ADR) and the revenue per available room (RevPAR) for the property.

These statistics are very useful for benchmarking the hotel's performance against other similar properties. However, they are merely statistics generated from the detail behind rooms revenue, which is detailed in the second section.

The second part of these summary accounts outlines the turnover (also referred to as total revenue), breaking it down into the constituent departments that generated the revenue. In the detail behind the summary sheet, the revenue stream will be further broken down, for example, the food and beverage (F&B) revenue will usually be broken down into the various areas where it is generated (i.e. 20 per cent of its food revenue from the bar, 15 per cent from room service with the remaining 65 per cent from the restaurant). However, this is likely to be further analysed so that the restaurant revenue can be seen to comprise 45 per cent breakfast revenue, 40 per cent evening revenue and 15 per cent for lunchtime.

The rooms department may, for example, break down the revenue by industry segment so that analysis can be made of where the most demand is being generated and what segments are the most lucrative in terms of ADR, or even by a geographic guest profile.

The third part of the accounts breaks down the departmental revenue into departmental profits. For example, the staffing costs in the restaurant along with the food costs will be deducted to show a departmental profit margin for the F&B department. Some hoteliers choose not to remove staff salaries when calculating the departmental profits which will lead to the profit margins looking much higher than would be expected if compared with other hotels. In this instance, staff salaries are usually charged against undistributed costs as a salaries and wages cost. However, allocating the cost this way is not strictly in accordance with USALI.

The fourth section comprises the undistributed operating costs which as the name suggests are costs that are accrued by the hotel that are not directly attributed to a specific hotel department. For example, one of the subsections is sales and marketing (S&M). This cost is essential to the success of the hotel but it cannot accurately be attributed to just one department, as good marketing may lead to a bedroom sale or a meeting room hire, but then the customer may also have breakfast, use the bar, spend some money in the shop, have a massage or

Table 9.1 Historic profits and loss accounts

Year	2014		2015		2016	
No. of rooms	150		150		150	
Rooms available	54,750		54,750		54,900	
Rooms sold	37,482		37,914		38,270	
Occupancy	68.5%		69.2%		69.7%	
ADR (£)	148.48		152.50		154.40	
RevPar (£)	101.71		105.61		107.63	
Growth (RevPAR)			3.8%		1.9%	
Revenues (£ '000s)						
Rooms	5,568.6	58.9%	5,781.9	61.1%	5,892.7	61.2%
Food and beverage	3,101.0	32.8%	2,867.3	30.3%	2,975.3	30.9%
Leisure club	586.2	6.2%	596.2	6.3%	577.7	6.0%
Other revenue	198.5	2.1%	217.6	2.3%	182.9	1.9%
Total revenue	9,454.3	100.0%	9,463.0	100.0%	9628.7	100.0%
Departmental costs (£ '000s)						
Rooms	1,536.9	27.6%	1,590.0	27.5%	1,644.1	27.9%
Food and beverage	1,885.4	60.8%	1,720.4	60.0%	1,782.2	59.9%
Leisure club	360.5	61.5%	378.6	63.5%	367.4	63.6%
Other revenue	68.5	34.5%	74.0	34.0%	69.5	38.0%
Total departmental costs	3,851.3	40.7%	3,763.0	39.8%	3,863.2	40.1%
Departmental profit (£ '000s)						
Rooms	4,031.6	72.4%	4,191.9	72.5%	4,248.7	72.1%
Food and beverage	1,215.6	39.2%	1,146.9	40.0%	1,193.1	40.1%
Leisure club	225.7	38.5%	217.6	36.5%	210.3	36.4%
Other revenue	130.0	65.5%	143.6	66.0%	113.4	62.0%
Total departmental profit	5,602.9	59.3%	5,700.0	60.2%	5,765.5	59.9%
Undistributed operating expenses (£ '000s)						
Administrative and general	775.2	8.2%	794.9	8.4%	818.4	8.5%
Sales and marketing	387.6	4.1%	293.4	3.1%	385.1	4.0%
Repairs and maintenance	302.5	3.2%	236.6	2.5%	288.9	3.0%
Utility costs	312.0	3.3%	312.3	3.3%	317.7	3.3%
Total undistributed expenses	1,777.4	18.8%	1,637.1	17.3%	1,810.2	18.8%
Income before fixed charges (£ '000s)	3,825.5	40.5%	4,062.9	42.9%	3,955.3	41.1%
Fixed charges (£ '000s)						
Reserve for renewals	0.0	0.0%	0.0	0.0%	0.0	0.0%
Property taxes	397.1	4.2%	406.9	4.3%	414.0	4.3%
Insurance	85.1	0.9%	94.6	1.0%	96.3	1.0%
Management fees	0.0	0.0%	0.0	0.0%	0.0	0.0%
Total fixed charges	482.2	5.1%	501.5	5.3%	510.3	5.3%
EBITDA (£ '000s)	**3,343.4**	**35.4%**	**3,561.4**	**37.6%**	**3,445.0**	**35.8%**

Figure 9.1 Typical bedroom

use the telephone. As such, the 'expense' may benefit all departments, or at least more than one department.

The last section is the fixed costs. These are costs that are borne by the hotel whether or not it is operating efficiently, for example, insurance costs are unlikely to decrease just because the hotel is running at 40 per cent occupancy rather than 90 per cent. Other deductions include any outgoing rent, property taxes and head office costs or any other fixed costs that are not specifically accounted for elsewhere.

There is also usually an FF&E allowance to ensure furniture, fittings and equipment can be replaced as and when required. The theory behind this allowance is that to enable a hotel to maintain its current level of trading, it needs to maintain the quality of the product and to do so it is usual practice to set aside a 'sinking fund' over and above the repairs and maintenance budget. That way, for example, the bedrooms can be systematically refurbished without needing to resort to capital expenditure. The amount that valuers will normally allow for this will be anywhere between 3 per cent and 5 per cent of turnover, depending on the quality of the property, its location and indeed the amount the market normally adjusts for a property of this type to budget for the FF&E account.

The fixed costs also have one other cost centre: management costs. If the hotel has a management cost it will be deducted here, and indeed in the event of the property being subject to a management contract with a base management fee and an incentive fee, then there will be two deductions.

However, a number of hotels are owner-operated and so do not deduct an allowance for the cost of management. In certain circumstances, it is appropriate

Figure 9.2 Hotel bar

for the investor to make a deduction to represent a notional management fee, for example, if the property is run independently at the moment but would usually be run by a chain operator who would either charge a management cost, or re-charge some central head office costs (like computer systems, human resources, accounting, etc.). In this case, it is essential that the investor (or his team of advisors) go through the accounts in detail to ensure that any costs that would not be accrued by a chain hotel are deducted so that the notional management fee is not 'double-charging' the hotel.

This will bring the hotel down to an EBITDA level, which is the standard unit that shows the level of profit that the hotel can make.

In many sets of accounts, depreciation and amortisation will be deducted leading to an EBITDA level, but as each 'hypothetical operator' will probably have different standards of depreciation and amortisation, as such they need to be added back in so that the profit margin is standard to all potential purchasers.

Other typical deductions that need 'adding back' are interest costs, directors' fees, legal fees and other costs that are being deducted that result directly from the operation of the current owner exclusively, and may be at different levels when the hotel is being operated by the reasonable efficient operator.

Unfortunately, it is not always the case that the investor will be presented with accounts that conform to USALI as in the example shown in Table 9.2. It is essential that the investor can still understand what the accounts are showing, and analyse them in sufficient detail to be able to ascertain the correct EBITDA levels.

Table 9.2 Historic profit and loss accounts – not in USALI

	2014	*2015*	*2016*
	£	£	£
Income			
Accommodation	244,811	252,155	259,720
Bar drinks	1,289	1,328	1,368
Breakfast	14,899	15,346	15,806
Interest receivable	68	70	72
Telephone/parking	515	530	546
Other	1,233	1,270	1,308
Total revenue	262,815	270,699	278,820
Cost of sales			
Commissions payable	10,865	11,191	11,527
Direct NI	3,565	3,672	3,782
Direct wages	75,357	77,618	79,946
Food and provisions	5,882	6,058	6,240
Laundry and cleaning	10,311	10,620	10,939
Purchases	491	506	521
Total direct costs	106,471	109,665	112,955
Expenses			
Accountancy fees	2,808	2,892	2,979
Advertising and PR	889	916	943
Bank charges	399	411	423
Credit card charges	1,965	2,024	2,085
Depreciation	7,545	7,771	8,004
Insurance	5,995	6,175	6,360
Interest – bank	7	7	7
Light and heat	6,525	6,721	6,922
Car	8,250	8,498	8,752
Directors salary	52,000	53,560	55,167
Rates	32,929	33,917	34,934
Rent	50,000	51,500	53,045
Repairs and maintenance	17,889	18,426	18,978
Reservations and book-keeping	2,150	2,215	2,281
Service charges	500	515	530
Software	225	232	239
Stationery and printing	130	134	138
Subscriptions	25	26	27
Sundry	1,875	1,931	1,989
Telephone and fax	2,366	2,437	2,510
Wages and salaries	18,500	19,055	19,627
Total expenses	197,881	219,361	225,942
Profit/(loss)	**(41,538)**	**(58,327)**	**(60,077)**

A number of items included as costs here to assess the tax position need to be added back to get to the sustainable EBITDA. According to USALI, the following costs would be struck out as shown in Table 9.3.

The investor should be satisfied as to the accuracy and reliability of the trading information and/or projections supplied for the purpose of the valuation. They should critically examine the details against either direct knowledge, or against the knowledge of their team, or of other hotels operating in similar trading circumstances.

The accuracy and reliability of trading information is generally established by the auditing process. The investor will need to be aware of the relative reliability of audited and un-audited figures, and also consider whether or not the accounts are qualified or are a preliminary report.

This is an incredibly important point: the accuracy of the historic accounts is usually the starting-off point for most investors when assessing the future potential of the property. Any suspected inaccuracies or defects should be questioned and answers should be sought from the current owner/operator.

Simple analysis of the profit and loss accounts

The investor will find it very useful to analyse the historic accounts that they have been provided with to start to understand the hotel operation. The following section outlines an example of the sort of analysis an investor might undertake when assessing a property's trading potential.

Section 1: the operating statistics

As can be seen from Table 9.4, the bedroom numbers stayed constant which suggests that no extension works were carried out during the three years that are being analysed. The rooms available in 2016 are higher than the following years because it was a leap year.

The occupancy rate increased marginally in 2016 compared with 2015, but 2015 showed a marked improvement on 2014 (almost 2 per cent). The investor would be asking themselves why this happened (was it an improvement in the general market, the result of improved demand at the hotel, change of management, capex investment in rooms or a change in market mix?) More importantly, the investor needs to establish whether it is sustainable. Would the new owner be running at close to 70 per cent occupancy or would 68.5 per cent be more likely?

There was also some growth in the ADR experienced over the years with a €2.01 increase in 2015 on the previous year (2.7 per cent) and a €1.90 increase in 2016 compared with the previous year (1.2 per cent). The investor will be considering whether the increases were purely down to inflation or market growth, or whether these are down to the specifics of the actual property. The lower level of growth in 2016 compared with the previous year will need explaining; has the performance of the property peaked in the previous year and

Table 9.3 Adjusted profit and loss accounts

	2014	2015	2016
	£	£	£
Income			
Accommodation	244,811	252,155	259,720
Bar drinks	1,289	1,328	1,368
Breakfast	14,899	15,346	15,806
Interest receivable	68	70	72
Telephone/parking	515	530	546
Other	1,233	1,270	1,308
Total revenue	262,815	270,699	278,820
Cost of sales			
Commissions payable	10,865	11,191	11,527
Direct NI	3,565	3,672	3,782
Direct wages	55,000	56,650	58,350
Food and provisions	5,882	6,058	6,240
Laundry and cleaning	10,311	10,620	10,939
Purchases	491	506	521
Total direct costs	86,114	88,697	91,359
Expenses			
Accountancy fees	2,808	2,892	2,979
Advertising and PR	889	916	943
Bank charges	399	411	423
Credit card charges	1,965	2,024	2,085
Depreciation			
Insurance	5,995	6,175	6,360
Interest – bank	7	7	7
Light and heat	6,525	6,721	6,922
Car			
Directors salary			
Rates	32,929	33,917	34,934
Rent	50,000	51,500	53,045
Repairs and maintenance	17,889	18,426	18,978
Reservations and book-keeping	2,150	2,215	2,281
Service charges	500	515	530
Software	225	232	239
Stationery and printing	130	134	138
Subscriptions	25	26	27
Sundry	1,875	1,931	1,989
Telephone and fax	2,366	2,437	2,510
Wages and salaries	18,500	19,055	19,627
Total expenses	145,176	149,534	154,017
Profit/(loss)	**31,524**	**32,468**	**33,445**

Table 9.4 Hotel key performance indicators – operating statistics

	2014	*2015*	*2016*
No. of rooms	150	150	150
Rooms available	54,750	54,750	54,900
Rooms sold	37,482	37,914	38,270
Occupancy	68.5%	69.2%	69.7%
ADR (£)	148.48	152.50	154.40
RevPar (£)	101.71	105.61	107.63
Change (RevPar)		3.8%	1.9%
TRevPAR (£)	172.68	172.84	175.39
Change (TRevPAR)		0.1%	1.5%

is now slowing down, or was last year affected by something that will not recur in following years?

Operators may decide to alter their business mix, either increasing occupancy levels at the expense of lower ADRs, or by increasing the ADR and lowering the volume of business through the property. This is called 'yielding' and it is the intention of every operator to try and optimise the rooms' yield that can be generated from the bedrooms by the best use of the mix of occupancy and ADR. There are arguments to say that an improved volume of business (at a lower rate) may lead to a bigger spin-off in secondary spending (with a higher volume of guests meaning a higher possibility of bar revenues increasing). There are also counter-arguments that say an increased volume of business leads to higher servicing costs and reduces margins, as well as increasing damage to the bedroom product through wear and tear.

It is the duty of the general manager to try and optimise EBITDA (or IBFC as many managers assume they can have limited impact on fixed costs) through efficient yielding.

To check the yield generated by the rooms, the hotel industry has come up with RevPAR (tevenue per available room), which in simplistic terms is usually the occupancy level multiplied by the ADR. TRevPAR (total revenue per available room) is an even better benchmarking statistic for comparing the hotel's current performance against historic levels as it analyses total revenue rather than just rooms' revenue. It is less useful for comparing with competitive hotels, unless they also have similar facilities to the subject hotel/resort.

Of course, the RevPAR (or TRevPAR) tool only reflects half of the story and the operator (and the investor) is also concerned with costs, so sometimes it can be useful to benchmark GOPPAR (gross operating profit per available room) to determine the profitability of the hotel.

As can be seen from Table 9.4, the RevPAR has been increasing each year, although the growth rate has dropped each year. Once again, the investor will be looking at this to try and understand what has been happening and what is likely to happen in the future.

Table 9.5 Hotel revenue by department

	2014 £ '000s		*2015* £ '000s		*2016* £ '000s	
Rooms	5,568.6	58.9%	5,781.9	61.1%	5,892.7	61.2%
Food and beverage	3,101.0	32.8%	2,867.3	30.3%	2,975.3	30.9%
Leisure club	586.2	6.2%	596.2	6.3%	577.7	6.0%
Other revenue	198.5	2.1%	217.6	2.3%	182.9	1.9%
Total revenue	9,454.3	100.0%	9,463.0	100.0%	9.628.7	100.0%

Table 9.6 Segmentation of rooms, demand rooms booked and occupancy

	2014 £		*2015* £		*2016* £	
Corporate individual	24,363	65.0%	25,782	68.0%	24,363	66.0%
Corporate groups	1,124	3.0%	1,517	4.0%	1,148	3.0%
Leisure individual	7,871	21.0%	7,204	19.0%	7,654	20.0%
Leisure groups	375	1.0%	375	1.0%	375	1.0%
Conference groups	3,748	10.0%	3,033	8.0%	3,827	10.0%
Total available rooms	37,482	68.5%	37,914	69.2%	38,270	69.7%

Revenues

The investor will look at the revenue mix of the property, once again trying to establish what has occurred historically to ascertain what is likely to occur in the future. Are there likely to be any key changes in the business mix or has a new facility opened or an old one closed, for example.

The detail behind the rooms' department accounts will probably analyse in fuller detail where the demand has come from, with the bookings divided into market segments. The accounts will probably detail the total revenue generated, the occupancy and the ADR for each segment, as shown in Table 9.6.

This will be extremely useful for the investor when they are looking at what is happening in the hotel as careful use of segmentation can enable the hotel to improve its performance.

Such analysis is also of benefit as it can outline if a hotel is too reliant upon any one segment. After the 9/11 terrorist attacks in New York, for example, a number of hotels across the world that had a high dependency on airline business found that they were either losing their business or that crew contracts were being renegotiated at lower rates, impacting quite substantially on their revenue generation. Any hotels that had a high reliance on such business found themselves badly affected compared with other hotels that had a better mix of room demand generators.

Other accounts may show the analysis of the top clients for the hotel, which will enable the investor to see how well-spread the business is at the hotel. Once

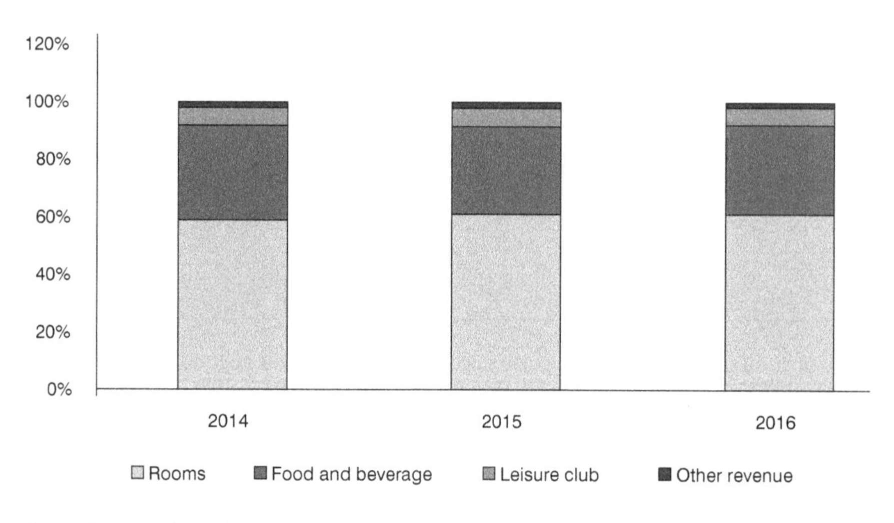

Figure 9.3 Analysis of revenue mix

again, too much business coming from one source could in certain circumstances be deemed 'at risk' from a change in the booking policy of that client in an economic downturn, and should be taken into account when undertaking the analysis of the accounts.

Looking at Table 9.7, the investor will see that although there was an improvement in rooms' revenue in 2015 on the previous year, total revenue only increased slightly.

F&B revenue was significantly down and the investor will want to understand why. Was a new menu introduced? Were the meeting rooms unavailable because of refurbishment? Were the number of weddings hosted by the property down on the previous year? Was an outlet (for example, a second or third bar or restaurant) closed down for the year? Did room service cease? Did a new restaurant open next door to the hotel? Has the allocation of breakfast revenue been altered in the way it is accounted for? The data for 2016 showed an improvement in the F&B revenue but it is still not up to 2014 levels; why?

Table 9.7 Departmental profits

	2014 £ '000s		*2015* £ '000s		*2016* £ '000s	
Rooms	4,031.6	72.4%	4,191.9	72.5%	4,248.7	72.1%
Food and beverage	1,215.6	39.2%	1,146.9	40.0%	1,193.1	40.1%
Leisure club	225.7	38.5%	217.6	36.5%	210.3	36.4%
Other revenue	130.0	65.5%	143.6	66.0%	113.4	62.0%
Total departmental profits	5,602.9	59.3%	5,700.0	60.2%	5,765.5	59.9%

The investor will be wondering why the leisure club revenue dropped down in 2016 after having improved in 2015. This may simply be down to a drop in membership numbers, a decline in membership monthly payments or possibly down to an inability to command the same joining fees. It could have come through the opening of new competition or because of an operational decision (perhaps the membership levels were too high and the use was adversely affecting the hotel guests' use of the facility).

The fluctuations of the 'other revenue' department will also be analysed to understand what has changed year on year, so the investor can try and predict what is likely to happen in the future. It may be that a significant proportion of other revenue comes from leased out areas, and that these could be terminating in the near future or could be due to be reviewed which could increase the rent received.

Departmental profits

As can be seen from Table 9.7, overall profits have remained relatively stable, with a slight improvement in 2015 followed by a slight downturn in 2016. The investor will look into why the rooms' department is marginally down over the period, checking whether it is a straightforward increase in staff costs, or whether there is something else behind the drop in profitability.

The improvement in F&B margins will need to be analysed, but probably goes some way to explaining why the drop in F&B revenue has been accepted by the management. The investor will be trying to determine whether there is any possibility that the margin can be further enhanced, whether the revenue can be improved from this department without affecting the profitability and indeed, whether the 40 per cent+ profit margin is actually sustainable.

The investor will also be keen to understand what has caused the margins at the leisure club to deteriorate, especially since it does not seem to be related to the revenue stream (sometimes an improvement in revenue can lead to an improvement in margins as the departmental fixed costs become more efficiently used).

The fluctuations in the profitability of the 'other revenue' will also have to be analysed to try and work out what is the likely position for the future, as shown in Table 9.8.

Undistributed operating expenses

Once again, the investor will try to analyse why the various subsections on the departments have been altering over the last three years to ascertain where the costs are likely to sit in the future.

One question raised by these accounts is that 2015's undistributed costs were significantly lower than those of either 2014 or 2016, and the investor will want to review these and see if they had an impact on revenue generation (specifically the drop in sales and marketing expenditure).

Table 9.8 Undistributed operating costs

Year	*2014* £ '000s		*2015* £ '000s		*2016* £ '000s	
Administrative and general	775.2	8.2%	794.9	8.4%	818.4	8.5%
Sales and marketing	387.6	4.1%	293.4	3.1%	385.1	4.0%
Repairs and maintenance	302.5	3.2%	236.6	2.5%	288.9	3.0%
Utility costs	312.0	3.3%	312.3	3.3%	317.7	3.3%
Total undistributed expenses	1,777.4	18.8%	1,637.1	17.3%	1,810.2	18.8%

Table 9.9 Undistributed operating costs – room analysis

	2014		*2015*		*2016*	
	PAR	*POR*	*PAR*	*POR*	*PAR*	*POR*
Administrative and general	14.16	20.68	14.52	20.97	14.91	21.39
Sales and marketing	7.08	10.34	5.36	7.74	7.02	10.06
Repairs and maintenance	5.53	8.07	4.32	6.24	5.26	7.55
Utility costs	5.70	8.32	5.70	8.24	5.79	8.30
Total	32.46	47.42	29.90	43.18	32.97	47.30

Another question would be the relatively static nature of the utility costs (energy costs) and how they have managed to remain a constant figure (in terms of proportion of total revenue) when the rest of the industry over this period is generally incurring higher costs.

The investor will also need to check what is included under repairs and maintenance (R&M) to ensure that all deductions are appropriate to be deducted under this section. Sometimes work that could be allocated as capital expenditure is mistakenly deducted here, which would adversely affect the EBITDA of the property.

It is usual for undistributed operating expenses to be analysed on a 'per occupied room' basis as well as a 'per available room' basis, so that these costs can be benchmarked against similar standard hotels in terms of efficiencies of operation.

In this example, the costs of administration and general (A&G) per available room has risen 5.3 per cent over two years, from €14.16 in 2014, to €14.52 in 2015 and €14.91 in 2016. On a per-occupied-room basis, the increase was lower, 3.4 per cent over the 2 years.

S&M costs dropped over the same period from €7.08 per available room in 2014, to €7.02 in 2016, with a larger decline per occupied room.

R&M costs have also declined over the same period, 4.9 per cent on available rooms and 6.4 per cent on an occupied room basis. The question to ask the general

Table 9.10 Fixed costs

Year	*2014* £ '000s		*2015* £ '000s		*2016* £ '000s	
Reserve for renewals	0.0	0.0%	0.0	0.0%	0.0	0.0%
Property taxes	397.1	4.2%	406.9	4.3%	414.0	4.3%
Insurance	85.1	0.9%	94.6	1.0%	96.3	1.0%
Management fees	0.0	0.0%	0.0	0.0%	0.0	0.0%
Total fixed charges	482.2	5.1%	501.5	5.3%	510.3	5.3%

manager would be why such a decline has taken place. Is the property genuinely needing less R&M or are some savings being made now that will require further expenditure in the future?

Utility costs on the other hand have shown a marginal increase on an available room basis (1.6 per cent), whilst declining (0.2 per cent) on a per-occupied-room basis.

Fixed costs

It is clear from the above that management have not specifically deducted anything for FF&E reserve, or indeed for a management fee or for head office costs. The investor will need to know if any head office recharges have been made elsewhere in the accounts, for example, a central marketing cost under S&M or a human resources recharge under A&G.

10 Valuation due diligence

Overview

The value of a hotel, whether annual or capital,[1] is based upon its potential trading, and in most instances an investor will buy (or rent) the property based upon the anticipated profit level it can sustain. For the potential investor, it is essential to know exactly how the property has been trading in the past to enable them to accurately assess how it is likely to trade in the future. This part of the due diligence process should always be carried out by an experienced hotel valuer. Comfort for the investor can be provided in this regard by using a RICS-qualified surveyor. The RICS has a registered valuer scheme that ensures each valuer is vetted for their level of experience, and RICS members cannot be 'registered valuers' unless they have the relevant experience.

It is important to remember that the hotel business can be highly cyclical, with trading in many categories of hotel highly dependent upon a number of outside events, including currency fluctuations, wars, interest rates, tourism cycles, terrorist acts as well as the more general economic cycles.

When carrying out the hotel valuation, getting to the sustainable EBITDA (or divisible balance in the case of annual valuations) is usually performed through five quite separate exercises:

- property/site inspection;
- analysis of the historic trading accounts;
- review of the competition;
- interview with the general manager/financial controller;
- independent analysis of everything that has been seen and said throughout the process to determine how the hotel is likely to trade in the hands of a reasonable efficient operator (REO).

In assessing the purchase price that can be afforded (or carrying out a market valuation), it is important that any 'personal goodwill' (extra trade that is generated by the vendor) associated with the property is ignored, whilst any 'property goodwill' (goodwill that stays with the property rather than the owner) associated with the property is included.

In assessing the potential performance of the REO, it is important to remove any element of 'personal goodwill' from the equation. As such, the valuer will be looking at the following points:

- whether the stated revenues and expenditures are in line with those considered achievable by a likely purchaser, or whether they are considered specific to the actual operator;
- companies with a number of outlets may not show outgoings such as marketing, training, accountancy, depreciation, cyclical repairs or head office management expenses in their individual unit management accounts;
- the accounts of an owner-occupier business may not outline management salaries or directors' remuneration;
- depreciation policies vary, and it is important that the valuer 'adds back' depreciation when assessing net maintainable profit;
- adjustments will need to be made to reflect the annual cost of repairing, maintaining and renewing the property and its fixtures, fittings, furniture, furnishings and equipment to an appropriate standard;
- hotels require refitting and re-equipping throughout their lifetime. It is important for the valuer to take into consideration the time, income, cost and impact upon profits of such events, especially if major capital investment is likely to be required in the near future. In this case, an amount may be deducted from the valuation to reflect this.

The inspection

The inspection provides one of the most important opportunities for the valuer to assess the value of the hotel. When reviewing a hotel, the inspection has five main functions over and above checking that the property actually exists.

1: Check existing facilities

To check the facilities contained within the property and to accurately audit what is and isn't included within the valuation. This may seem obvious, and would be undertaken by all valuers in an inspection of any type of property. We have highlighted it here as it is not always simple when dealing with hotels.

It could be that the property includes an annex with staff accommodation, that a conference building is held on a short lease (or different title than the main property) or that some of the plant and machinery items (for example, telephone switchboxes and boilers) are leased rather than owned, and this may only become evident during the inspection.

Complications can even arise as to how many rooms the hotel has, with some sources reporting in 'rooms' and others in 'keys'. Care must therefore be taken as a suite which has two bedrooms may be included within the letting inventory as 'two rooms', but could at the same time be referred to as 'one key'.

2: Assess the condition

To assess the size and condition of the property, to ensure it is in good condition or to enable the valuer to make any necessary allowances for future capital expenditure. It is important that the condition of the property (with or without capital expenditure) ties in with the forecasted projections.

For example, a hotel may be able to achieve a RevPAR of €165.00 if it is in good condition, but if the property has slightly more basic rooms and facilities, it may only be able to achieve a RevPAR of €130.00, and without a full inspection, it is unlikely that the valuer will be able to ascertain the probable trading level.

Alternatively, the hotel may be projecting an ADR of €140.00, but it may contain too many single rooms to achieve such rates. Then again, it might be showing minimal conferencing revenue (at a lower level than would have been anticipated considering the listed facilities) and the reason for such a poor trading record could become evident during the inspection.

3: Analyse financial implications of the layout

The inspection will allow for a careful analysis of any revenue and cost implications from the layout of the hotel. The design of a property can affect the efficiency of the hotel operator and it is much easier for an experienced valuer to analyse how the property works during an inspection, rather than just from floor plans.

For example, limited-service hotels strive to provide value for money accommodation to a relatively price-sensitive client base, with customers generally keen to stay in the most competitively priced property of a certain standard. As such, it can be important to minimise undue costs, including an effective use of staff. A number of budget hotels have therefore 'designed in' operational savings and these will impact upon the EBITDA potential of the property.

A specific example of this type of design is contained with the typical Holiday Inn Express floor plans. The reception area (which needs to be manned throughout the day and night) is adjacent to the bar area (which needs less manning during the day). The idea is that during the day (and during quiet evenings) the reception staff can also provide a bar service, enabling the hotel to operate this facility without having to man both areas independently, thereby increasing potential revenue generation without incurring additional staffing costs.

The design and layout of the hotel's facilities can have an impact upon their revenue operating potential. Meeting rooms without natural daylight can sometimes be difficult to let. Bars and restaurants with separate off-street access could potentially attract significant non-residential custom. If the vast proportion of bedrooms have excellent views, they could potentially command an enhanced rate.

4: Look for treasure hunting opportunities

It is considered a truism that in any business, the longer you have been with a specific organisation, the less aware you are of other ways of doing business, or to put in into other words, the less able you are to see unexplored opportunities. It can become common for a hotel to concentrate on what it is doing and only improve the aspects of trading that have been determined to be the priorities, sometimes to the detriment of other potential trading opportunities.

Other items that can be potentially 'undervalued' in hotel valuations include unused or spare land that does not generate any income for the hotel, and could be sold off without impact on the operation of the hotel.

Staff accommodation that is owned but no longer needed for the operation of the hotel can sometimes add significantly to the overall value of the hotel. For example, if the accommodation can be sold off independently, or converted into revenue-generating guest accommodation, it can enhance values. However, it is very important to ensure that the true cost implications are taken into account; staffing costs may be lower because of the provision of staff housing and as such, an allowance would have to be made to reflect this.

It is not the case that all staff accommodation is unimportant to the operation of a hotel and can therefore be sold off. If staff cannot be attracted to work at the hotel without the provision of staff accommodation, then it must be accounted for as part of the operation of the business rather than as a separate element of the property.

Other areas of potential treasure hunting include hotels that are not optimising the use of their facilities (bedrooms could be converted into meeting rooms, free city centre car parking when it could charge for such facilities). All of these need to be explored to assess what the REO could achieve.

Since it is part of the valuer's duty to look for such treasure hunting opportunities, anything that is discovered, even if not something that is likely to be undertaken by the REO, will usually be mentioned in the valuation report.

5: *Review alternative uses*

The valuer needs to look at the value of the hotel and review whether an alternative use would provide a higher value. This can be very important because a number of rundown hotels with poor trading records could have substantially enhanced values for alternative uses than those that a hotel operator would be prepared to pay for the property.

There are a number of common reasons why a hotel could have a higher alternative use value, including the following:

- it is poorly located for the use as a hotel (perhaps it now has diminished access after development around the property);
- the property no longer meets the requirements of its marketplace (for example, no en suite bedrooms);
- the property has been left to deteriorate and would cost too much to bring it back into a good enough condition to trade in its market;
- the property was poorly located when built and was only constructed as a planning condition of a larger mixed-use scheme;

- other land uses in the area have increased at a faster rate than hotel land values.

When reviewing alternative use value, the valuer must be aware of the local planning regime and take into account the uncertainty of gaining planning permission for a new use, as well as the cost of demolition, conversion and rebuilding.

Limitations of the inspection

The problem with inspecting a hotel is that it is normally impossible to visit every bedroom during one inspection, because a number of the bedrooms are likely (well, hopefully) to be occupied by guests at any one time. That said, it is not usually necessary to inspect every bedroom, although only the visiting valuer will be able to make that determination. If the property is a 150-bedroom chain hotel and all of the rooms are of a similar size and condition, then less rooms will need to be seen compared to a hotel that has very few rooms of a similar style, size or quality. It is important to bear in mind the objective of the inspection when determining how large a sample the valuer needs to review. The valuer needs to be able to review the historic trading and future projections accurately and to do this, a reasonable cross-section of bedroom accommodation will need to be viewed, including the best and worst rooms.

Sometimes it may not be in the vendor's interest to show the least attractive or smallest rooms to the valuer and they may try and hide certain areas; during the inspection, it will normally become apparent to an experienced valuer if this is happening and it is normal practice for the valuer to then start to ask to see specific rooms and require a much more comprehensive inspection.

Time of the inspection

The time that the inspection is carried out is likely to affect the number of bedrooms that a valuer is able to see. Peak guest times (early morning and early evening) are the least desirable from the valuer's perspective, as the least amount of rooms will be available for inspection.

However, if the majority of the revenue comes from the meeting rooms, an early inspection time may be preferable so that this area of the hotel's accommodation can be fully inspected before it starts to be fully utilised. Alternatively, if most of the revenue comes from the bar or restaurant, the valuer may wish to inspect the property whilst these areas are in operation so the pattern of trading can be assessed.

Ideally the valuer should have the opportunity to stay overnight at the property thereby experiencing the property as the guest would, prior to formally inspecting the property and interviewing the general manager.

It is important that the inspection continues until the valuer is happy that they have seen a good cross-section of the hotel's trading facilities and back of house areas (areas used by staff for the operation of the hotel, rather than being used by guests and directly generating revenue), as these will impact on operational costs and capital expenditure requirements.

It will be impossible to judge the correct EBITDA without having seen a satisfactory cross-section of the hotel, and if the first inspection is not sufficient to provide an adequate sample, then another inspection should be carried out.

The local hotel market

It is vital for the valuer to understand the specific parameters of the local hotel market if an accurate assessment of the sustainable EBITDA is to be possible. In that regard, the valuer must find a way to assess the competitive supply of hotels in the area, the demand generators, and any proposed changes to the competitive supply.

The valuer should also try and determine the usual performance in the area through historic benchmarking statistics as well as understanding the underlying nature of the hotel market.

The competitive supply of hotels

The first key question is 'what constitutes competition for the subject hotel?' The answer to this will vary for every single property. A hotel will provide competition for another hotel if they compete in any single aspect, whether in food and beverage trade, leisure club guests, conference room demand or for room business. That said, it is usual for a hotel's main competition to come from hotels of a similar quality in the local vicinity that compete for similar clientele. For example, in London the Four Seasons Canary Wharf is located less than 1 mile from the Ibis Docklands and they do not provide any competition for each other, in any segment of business.

This example is perhaps slightly exaggerated because the difference in the quality of the products (the Four Seasons being a deluxe five-star property and the Ibis providing budget accommodation) and the differences in the markets in which they compete means that there is very little potential crossover of potential trade.

However, it is quite possible for the supply of budget hotels to have an adverse impact upon higher-grade hotels, and in many areas the mid-market hotels have seen their profitability decline after new openings of budget hotels in the same area.

Which hotels provide direct competition will normally be determined by the existing revenue profile, although it is essential for valuers to assess how the hotel could trade in the future and review the competition (and potential competition) accordingly.

Competition will not just come from other hotels. If a high proportion of revenue comes from the restaurant and bar area, then other restaurants and bars need to be considered as competition. If the hotel has a fitness centre with an external membership, then other health clubs will provide direct competition to the hotel.

Competition can also come from properties outside of the immediate area. For example, top quality country house hotels that generate their own demand (i.e. not location based) are likely to compete with other such properties, even when they are some distance from each other.

Another example is when a high proportion of revenue for the hotel comes from coach tours or other forms of group tourism. In such cases other hotels that compete (or try to compete) in such markets should be reviewed, to see whether the current revenue generated from this source is likely to continue unchanged.

It would be usual for a valuer to inspect the most important competition, usually by requesting a guest show-round from the reception or meetings and events (M&E) staff. Hotels are regularly asked to see the quality of their bedrooms and meeting room facilities, and will normally try to accommodate such requests. It must be borne in mind that it is usual to show only the most recently upgraded bedrooms and as such, any rooms that are seen may not be truly typical of the hotel's bedroom supply.

A number of hotels also provide details of their rack rates (the published room tariff). It will be useful to note these, although the usefulness of these rates as an assessment as to what a guest is expected to pay, will depend upon the pricing policy of the hotel. Some hotels will quote a rack rate and will expect to achieve that rate (this is more usual of the limited-service hotel groups who operate fixed pricing policies), but a large number of hotels show an inflated 'rack rate' and then offer substantially discounted rates that render the disclosed rate almost irrelevant.

Rack rates

Trevor Ward, managing director – W Hospitality Group

Rack rate? How so? What rack?!

Well, in the days before computers, we had to manage the letting of the rooms manually. We had to look at books, into which the reservations agent would write in guests' names, and there was lots of rubbing out and rewriting. And they would write out, for each reservation, a reservations slip, in multiple copies, one of which would fit into a slot, one for each room, on something called a Whitney Rack.

This was a board with tiered slots, one for each room, into which that reservation slip would fit, plus a coloured piece of transparent plastic, different colours denoting the status of the room – green for available, red for let, yellow for out-of-service, and so on.

So that's why they call it 'rack rate'!

Future competition

If the supply of hotels in a particular market segment changes, it is likely to have an impact upon the trading potential of the subject hotel. The closing down of competitive hotels can lead to increased demand for bedrooms in other local hotels, and so it is important to try and ascertain if any hotels are expected to close down in the next few years. This can be quite difficult, especially if the competition comprises mainly unbranded independent hotels.

It is generally much easier for the valuer to discover proposed new developments, although determining which schemes are likely to come to fruition can prove difficult, and will rely upon the experience of the valuer and the quality of the information provided.

The demand profile of the area and potential changes in demand

The performance of a hotel will be influenced by the level of demand in an area and the nature of that demand. For example, hotels in Kokand, Uzbekistan will be influenced by a number of demand generators, including its position as a regional commercial centre for the Fergana Valley which will generate strong national, regional and local corporate demand for hotel bedrooms during the mid-week period. It will also be strengthened through good tourist demand (mainly from nationals rather than international visitors), which will strengthen some of the quieter corporate trading periods as well as generating bedroom demand over the weekends.

To determine the demand generators in an area, the valuer will need to review both the business and the leisure profile of the location to determine where the trade is likely to come from. If the location is a relatively poor area with high unemployment, this may suggest that the corporate base for a hotel may be quite poor and so the emphasis behind the leisure trade will be more important.

Unfortunately, it is not quite as easy as just counting the local businesses as certain industries generate more bedroom demand than others. For example, banking generates more demand for hotel bedrooms (through training courses, relocations, etc.) than farming generally does. It is also the case that different companies will require different hotel accommodation, so top accountancy firms may have more need for top quality hotels (to accommodate their clients, partners, etc.) than a mechanic training school where budget accommodation is more likely to be required.

Leisure tourism is no easier to determine accurately, as only certain countries and cities appear to accurately account for overnight visitors to an area. The popularity of a visitor attraction will not necessarily be converted into demand for hotels, as some attractions are not final destinations and do not therefore attract accommodation requirements as often as the visitor numbers would suggest. For example, the Palace of Versailles is one of the most popular tourist attractions in France. Unfortunately, because its location is so close to Paris, it is usually a day trip or a stop-off on the way to another destination (rather than

Figure 10.1 Hotel bedroom – key driver behind revenue generation

Figure 10.2 Design of reception and bar allows for staff cost savings

being an end attraction, as such) and so does not generate the demand for hotel accommodation that would otherwise have been expected from such high visitor demand.

Once the demand profile has been successfully identified, it will be important to look at the dynamics of the demand and review if this is likely to change. Changes can come from many areas, including an increase in new companies moving into the area, change in the economic profile of the area, changes in transportation or changes in inward investment.

Performance benchmarking

It should always be remembered that although a valuer is looking to value the performance potential of a unique hotel, the value is based on how the REO would be able to perform. If the hotel is underperforming or over performing at a level that the hypothetical operator could not be expected to attain, then this is ignored when assessing value. It may well be that the prospective purchaser would perform at an 'atypical level', but that is irrelevant in value terms.

This does not mean, however, that the performance should be the average performance of all the potential operators for the unit. Some hotels will attract either particularly high-performing REOs, and others will attract those that do not perform at such high levels.

To be able to accurately ascertain how well (or poorly) the hotel is performing, it is useful to benchmark the hotel's performance against its competitive set. The valuer will usually have enough immediate experience in the particular location and market segment (and indeed contacts) to know how that particular market set is performing.

More general information on performance levels is available from companies such as STR Global who provide either generic location information (for example, all chain hotels in Moscow ran at an occupancy level of 69.2 per cent at an ADR of €188.95 in 2014), or more detailed performance information (for example, the average occupancy level of five specified luxury hotels in the Kitay Gorod district of central Moscow that can be used as a competitive set benchmark).

Although such benchmarking will not determine what level of occupancy or ADR the valuer should attribute to the subject property, it is exceedingly useful in assessing the various underlying trends in that local market.

It is also useful for the client (whether they are a bank looking to lend on a property, a potential purchaser looking to buy a property, the board of directors in an annual review or an operator looking to take out a lease on the property) to see how the property has been comparing year on year with the local market.

The interview and analysis

The purpose of the interview with the general manager is to determine how the property has been trading historically, and the likely trading in the future. Ideally, it is useful to undertake the interview face to face and after the accounts have

The benefits of benchmarking

Thomas Emanuel, director of business development – STR Global

How would a sport team be able to evaluate their own performance without playing and comparing themselves against the others? Competition represents a powerful motivational force for improvement and success. Through competition, our strengths and weaknesses are revealed and winning ideas are inspired. In business, the practice of measuring an organisation's performance against their peers' is known as 'benchmarking'.

Hotel benchmarking enables a property to gain knowledge of its competitive position versus other properties in the same market, the whole market, or even another market. The knowledge gained helps identify the performance gaps based on established facts and opens a great opportunity for continuous improvement.

You may ask yourself whether the rate strategy should be reviewed if your hotel's average rate is lower than the average of your competitors. Or, you may wonder whether your weekend marketing strategy is encouraging higher occupancies on weekends in comparison to the non-promoting properties. A benchmarking report may tell you valuable things about your property and thus, inform important strategic decisions.

STR Global provides hotel data and benchmarking reports to hotel operators, developers, financiers, analysts and suppliers to the hotel industry; covering daily and monthly performance data, forecasts, annual profitability, food & beverage, pipeline and census information worldwide.

It is an invaluable tool for all people involved in the hotel industry, and the careful choice of benchmarking data is essential in understanding the strengths and weaknesses of any particular hotel or resort.

been received, as some obvious questions may arise from the accounts that may not otherwise be addressed automatically during the interview.

It is normal to send out a wish list of information prior to inspecting the hotel that will hopefully be received prior to the inspection as these items are likely to provide a good basis for the interview questions.

The wish list may include the following items.

Legal information

- title documentation or reports on title
- ground leases or property leases
- subleases
- occupational licences or concessions
- details of any problems with the title

- planning consents for each property
- details of any unused planning consents
- details of any unfulfilled planning conditions
- copies of the operating licences
- copy of the management agreement
- copy of the franchise agreement.

Property information

- areas of the constituent parts of the property
- site plans (with sizes)
- building plans – floor by floor
- breakdown of bedrooms, F&B outlets, etc. by number and size
- copies of any current or historic structural surveys
- copies of any current or historic valuation reports
- copies of any current or historic condition surveys, asbestos surveys, etc.
- details of historic capital expenditure on the properties (for the last three years)
- details of any proposed capital expenditure (current year and next year's budget).

Operational information

- historic profit and loss accounts (three years minimum; this will be the full P&L accounts, not just the summary sheet)
- statutory accounts
- full profit and loss budget for the current year and for the next three years (if possible)
- details of market segmentation and geographical analysis of the guest profile
- details of contracts to supply bedrooms both historic and ongoing (corporate contracts/airline contracts, etc.)
- staff details – numbers of staff and contracts of employment, including pension details, length of service and skills training (including languages) for each member of staff
- sales and marketing plans; this year and budget next year
- details of food and beverage revenues, where it comes from, average cover spends, sleeper:diner ratios, etc.
- details of the local markets; SWOT analysis of the local competitors
- details of local trading; how does the property compete with its competitive set.

Other

- details of any outstanding complaints, legal claims, etc.
- details of the tax positioning of the property (outstanding tax bills, capital allowances, etc.)

It would be impossible to write a list of all the questions the valuer needs to ask during the interview, as the answers that are received will lead to additional questions until a full understanding of the hotel's operation is gained.

It is considered useful to go through the interview in a logical way, perhaps following a similar approach each time to ensure that nothing is forgotten.

One such approach would be to review the trading in a similar way to the layout of the accounts, for example, going through each departmental profit line first and then moving down to the EBITDA. So for example, in a 100-bedroom conference hotel with leisure facilities, the question structure might be something similar to the following.

Bedrooms

Historic accounts – first the valuer will try to understand the past performance of the hotel:

- what was the historic occupancy level/ADR?
- were the historic levels typical of the market?
- was there anything in particular that influenced the result (either positively or negatively)?
- what was the segmentation of the business like – how many of each type of customer paying what different rates?
- who were the top ten clients for the hotel and what proportion of total revenue did they generate?
- was there any specific geographic orientation to the clientele?
- has any capex been spent on the bedrooms recently, or do they need capex?
- what profit margin is the rooms department running at?
- is housekeeping outsourced, is the head house-keeper experienced and a long-term hotel employee?

Then the valuer will try to understand the potential future trading profile for the property:

- what are the budgeted occupancy levels and ADR?
- are they achievable? (sometimes the budget is fixed by others, or is set at a very challenging rather than realistic level)
- are they reliant upon proposed capex?
- how will they be achieved?
- how do they compare with the competition?
- are they predicting any change in segmentation?
- is there new supply coming online (or hotels closing down) that will affect future performance?
- are there any major changes to the demand for such hotels in the area?
- what proportion of the budgeted trade is reliant upon one area/company/demand generator that could be at risk?

- are there any changes that can be made to the cost structure of the department that will not adversely affect the quality of the operation?

The idea of these questions is to determine how the hotel rooms department functioned, to gain an understanding of the demands and opportunities that exist for that segment of the hotel. It may be that each question leads to many more on a similar vein until the valuer understands how that department works.

The same process will be undertaken with the F&B department.

Food and beverage revenue

- where did the revenue come from (in terms of bar, restaurant, room service, banqueting, etc.)?
- is there any opportunity to change the revenue mix between the departments to improve both revenue generation and profitability?
- what was the breakdown in revenue between breakfast, lunch, afternoon tea, dinner, etc.?
- what were the guest conversion rates like (the proportion of guests that stay in the hotel to have dinner)?
- who are the local competitors for lunchtime/evening meals and how do they compare with the hotel's offering?
- is there much demand from non-resident guests for the bar or restaurant?
- what opportunities exist to expand the F&B revenue?
- how many weddings does the property cater for a year and can this be improved?
- what is the volume of business like in the meeting rooms? Are day-lets prevalent or is it easy to sell day delegate rates and residential packages to the guests?
- are the rooms the right size for the market and do they provide the relevant level of flexibility?
- are there any areas where the profit margins for the F&B are particularly low? How can these be improved?
- what level of service do the guests demand?

The valuer should be aware of the different stresses that come from running the F&B operations in a hotel, and understanding the peculiarities of cost control in this department will be essential. Almost every sub-department, whether the bar, the restaurant, the room service, the meeting and events, will have individual idiosyncrasies that will affect their potential profitability.

It may be that the REO would consider it sensible (and more profitable) to lease out the restaurant to a third party, as a higher level of rent would be received than the hotel generates in way of departmental profit.

Then the valuer would start to ask similar questions on the other revenue lines (including the leisure club revenue, car parking revenue, telephone revenue and minor operating department's revenue). By working through each department's historic trading and future potential, the valuer should be confident of knowing precisely where they feel the average operator could expect to trade in terms of total revenue and departmental profits.

Hotels need mice!

Matt Weihs, managing director – Bench Events

MICE, the general term for meetings, incentives, conferences and events, is a term generally shunned by the industry but it does serve the purpose of categorising the different types of products a hotel/conference space services. The industry, on the whole, would prefer the broader term 'business conferences' to MICE and, across Africa, it is this phrase that accurately describes the real possibilities open in the conference market across the continent.

Conferences and events provide large hotel and travel receipts. Business travellers increase occupancy, demand for F&B and, if an event becomes an annual fixture, provides steady forecasts and stable revenues for a hotel. Across Africa, a majority of travellers are business related so it would serve a hotel well to entice bigger groups such as conferences. However, in emerging economies, events are also used to create societal impact. This is more than providing direct jobs from travel and tourism which, of course, they do. This is more about using conferences and events to educate a market of the potential of investment. And it is to this end that many of the African governments are leaning towards the conference industry. They are, therefore, making their own investments to attract the right types of conferences to their cities. This creates more travel receipts, underpinning tourism, and will increase jobs but here is an opportunity to impress upon delegates the incentives for investment and hold the attention of the next wave of investment.

One of the annual challenges for the leading Hotel Investment conference, hosted by Bench Events, is actually finding a relevant venue to host it. There is an irony that, for the principal annual conference for the hotel development industry in Sub-Saharan Africa, there is a lack of real choice when it comes to locations that have suitable facilities for such an event. Some markets have no choice at all but certainly all, with a lack of real industry all have a reduced 'on-the-ground' services available for the event. This lack of competition and high energy costs make setting up a conference quite a costly and therefore risky proposition. However, the pipeline of new conference facilities, looks very exciting and in this respect – with many state of the art conferencing facilities coming online in the near future – it will make the market much more competitive.

This questioning will then go on to discuss the undistributed operating costs and fixed costs until the historic details can be understood and the future predicted with some accuracy.

However, sometimes it is not possible to interview the general manager or financial controller through absence, or the length of time for the interview is too short to really get a feel for the operation of the hotel. Other times the GM may be inexperienced or the valuer may be expressly forbidden from carrying out an interview. In these instances, the valuer will only be able to get a partial feel for the operation of the hotel, in which case a more detailed inspection and further local enquiries may be required to enable them to feel confident enough to produce a shadow P&L account for the hotel.

At the end of the inspection, having gone through the accounts and undertaken the interview with the general manager, the valuer should feel relatively confident about completing the shadow profit and loss account for the hotel, at least up until the IBFC line. It may be that the projections cannot be completed to the EBITDA line until the other enquiries have been completed.

Brand names

Should the property valuation include a brand name (or the rights to a trading name) then the potential of that name should be included. This value will form part of the 'property goodwill' rather than the 'personal goodwill' attributed to the property. If in the valuer's opinion, a prospective purchaser bidding for the property in the market might consider that the hotel, if sold without the brand name, may trade at a different level then the difference in value should be identified.

Likewise, where a hotel without an existing brand or trading name is likely to be purchased by an operator with the benefit of a brand (bearing in mind the REO), then the increase or change in maintainable earnings and resultant value will need to be reflected by the valuer.

Independent analysis of the information

The final stage of the process is vitally important. The valuer will have a good understanding of the individual dynamics of the hotel and will know what facilities they have, what the local market is like, where the hotel has been (in terms of trading), where it hopes to go and how it hopes to get there.

The valuer will put all of this together and draw upon their knowledge of other similar hotels in the local area, as well as other comparable locations to assess what the market would pay for this specific hotel.

The valuer would then compare the potential performance of the subject property against the potential performance of hotels that have recently sold, so that an accurate projection of trading for the REO can be determined.

The valuer will draw upon comparable evidence to compare:

- RevPAR projections with the market
- revenue mix with the market
- departmental costs with the market
- undistributed costs with the market
- fixed costs with the market
- capitalisation rates or discount rates with recent transactions
- sales prices on per bedroom basis with recent transactions.

Problems with comparable evidence

Ideally, the valuer will be able to refer to market evidence to determine the appropriate multiple to apply to the EBITDA. The best evidence of what yield to use comes directly from recent market evidence of comparable properties, although there are problems with comparable evidence, which we outline below.

Accuracy of the information provided and lack of transparency in the hotel transactional market place

The hotel property market is a relatively secretive world and transactional information is rarely available to the general public. Even when the price paid is accurately reported (and many times it is not accurately reported), it is highly unlikely that EBITDA numbers and trading projections will be fully available to accurately analyse the multiplier that was applied.

There is very definitely a ranking system for comparable evidence in which first-hand, detailed transactional knowledge is the most desirable sort of evidence, moving down the scale to partial hearsay evidence.

A valuer should be aware of the most comparable transactions in the market and hopefully will have access to the yields in transactions so that they can be satisfied that their valuation is reflecting the current market.

Comparable evidence that the valuer does not have first-hand knowledge of needs to be verified. Partial hearsay evidence can only ever be used as the broadest of all benchmarking tools, to set the tone of the market, rather than provide direct evidence.

However, care must always be taken as partial information can be quite misleading as is shown in the text box on page 203.

Comparability of the transactions

There are grave difficulties with the use of comparable evidence in the hotel world since very rarely are there two hotels of a similar enough nature and trading profile to warrant anything other than using the transaction as a benchmark, rather than hard evidence.

Tsogo Sun – analysing transactions accurately

In 2013, one of the more significant single-asset hotel transactions in Sub-Saharan Africa occurred, when Tsogo Sun took ownership of the Southern Sun in Ikoyi, Lagos, Nigeria for a reported $70 million.

The first reports suggested that the transaction of this 195-bedroom hotel broke down to a sale price of $358,975 per bedroom, but the full information had not been taken into account and this information was inaccurate. The $70m was only for a 75 per cent share in the hotel but it also included a piece of land for redevelopment. As such, any analysis needs to take into account the full details.

In addition, it is usual to review sales prices based on a yield basis, and the first reports suggested that the hotel had transacted at 7.85 per cent (this is calculated by taking the net income (EBITDA) and dividing it by the gross transaction costs (price paid and fees). However, this analysis had taken the net operating income discussed in the sale memorandum at face value, and had not deducted the necessary costs to bring it down to an accurate EBITDA.

Our analysis of the transaction suggests a price paid of $341,880 per bedroom and a yield of 11.35 per cent, although this is slightly dependant on the amount of value attached to the land in the Golden Gate complex that was included in the transaction.

In addition, there is rarely a good supply of comparable evidence to enable the valuer to be able to identify the perfect comparable transaction. Most locations do not have large stocks of hotels and therefore any local transaction tends to be reviewed as a potential comparable property.

It is also common for valuers to look through comparable evidence further afield if the property type is comparable. For example, a four-star hotel in San Pedro Sula, Honduras may sometimes be referred to by a valuer looking for comparables for a similar hotel in Tegucigalpa, even though the dynamics of the two markets are quite different.

It is therefore common for the comparable evidence to be adjusted (through the valuer's market knowledge and experience) to make it comparable to the subject property.

Analysis of the comparable evidence and the treatment of costs

It is important for the valuer to know what they are looking at when they are consulting comparable evidence. For example, the press may report, for example, that a particular hotel sold off an 8 per cent yield and the valuer will need to analyse the details behind the transaction to see if the reports were correct.

Many times, the reported earnings may not be accurately reported, but assuming that the EBITDA is correctly reported then the valuer must look to see how accurate it is. For example, have management costs and an FF&E reserve been deducted in the reported EBITDA?

Example: Hotel Gilchrist

In this example, the press has reported that a 100-bedroom hotel sold for $10,000,000 (including costs) of an EBITDA of $1,000,000, suggesting a 10 per cent capitalisation rate.

Purchase price per bedroon	$1,000,000
Yield	10%

It is subsequently learnt that the turnover was $3,000,000 and that a management fee and FF&E reserve had not been deducted when calculating the reported EBITDA.

It is deemed appropriate that a 3 per cent management fee is deducted along with a 4 per cent FF&E reserve for this type of property.

Turnover	$3,000,000	
Reported EBITDA	$1,000,000	
Less FF&E reserve at 4 per cent	$120,000	
Less management fee at 3 per cent	$90,000	
Adjusted EBITDA	$790,000	
Sales price (including costs)	$10,000,000	
Yield		79%

Capitalisation rate

It is not always appropriate to deduct management costs or FF&E reserves when carrying out a valuation. The valuer will reflect what the market does when assessing the value of the property. If it is inappropriate to deduct a management fee or an FF&E reserve, then the valuer should ensure that the comparables they refer to have been calculated on a similar basis.

Determining the appropriate yield to apply in a hotel valuation is an extremely complex business, and requires both experience and full exposure to the current market. All relevant comparable transactions must be fully analysed although because of the individual nature of the business, it is likely that many adjustments will be required (or at the very least considered) to enable the value to correctly reflect the markets perception of the appropriate yield.

Note

1 A capital value is the value of the entire interest of a property, whether the freehold or the leasehold interest. An annual valuation is the value of the same property for one year, in effect its rental value.

Part III
Specialised valuation categories

Specific examples

The final section provides examples of specific types of valuations. The details from Part II on due diligence apply to all of these individual chapters, but each section discusses specific issues arising from that type of property.

There are individual chapters on spas, gyms and golf courses, all of which can be stand-alone leisure properties or can form part of a hotel or resort.

In addition, other chapters cover serviced apartments, fractional ownership, site valuation and rental valuations.

11 Rental valuations

Undertaking a rental valuation (an annual valuation) for a hotel is similar in many ways to valuing the property on a capital basis. The methodology that should be used is dictated by the local market, but is generally either the profits method or the comparison method, or more usually, a combination of the both.

The 'Red Book' defines the rental value of a hotel as 'the estimated amount for which an interest in real property should be leased on the valuation date between a willing lessor and a willing lessee on appropriate lease terms in an arm's length transaction, after proper marketing and where the parties had each acted knowledgeably, prudently and without compulsion.'

As such the duty of the valuer is to determine what any potential tenants in the current marketplace would be prepared to pay. It is therefore essential that the valuer reflects the approach adopted by that same market when deciding which methodology to employ to calculate the market rent.

As can be seen, the definition of market rent is a modified definition of market value hence the immediate similarity. The market rent may vary significantly depending on the terms of the assumed lease. In the event that there is not an existing lease, the appropriate assumed lease terms will normally reflect current practice in the market in which the property is situated, although for certain purposes unusual terms may need to be stipulated. Matters such as the duration of the lease, the frequency of rent reviews and the responsibilities of the parties for maintenance and outgoings will all impact the market rent.

In certain countries or states, statutory factors may either restrict the terms that may be agreed, or influence the impact of terms in the contract. These need to be taken into account wherever appropriate.

In most instances the agreed rent for a property is exclusively determined by the potential profitability of that type of hotel, albeit that the potential level of trade will be assessed using market evidence. It is not solely determined by comparable evidence for other rental agreements as can be usual with other property classes. Market comparables should always be referred to, to see whether the end result appears reasonable.

However as stated earlier, it is essential the valuer reflects the local market. If the location and type of property is such that rents are set by reference exclusively to market comparables, then that is the approach the hotel valuer should take.

Hotel rent review – comparable transactions

Many types of commercial properties, such as offices, retail or industrial, are occupied under short leases which enjoy/suffer (depending on your point of view) rent reviews at regular intervals.

On the other hand, a lot of hotels are occupied/operated either freehold, or under long leases, or under management contracts where rent reviews do not occur. As such, rent reviews in the hotel sector are relatively scarce compared with other sectors of the property market.

However, that does not mean they do not occur. In the limited-service sector, where occupiers such as Travelodge or Premier Inn operate under fifteen to thirty-five year leases, rent reviews are quite common. Some of these rent reviews are subject to a formula – for example, the change in RPI/CPI – but others are assessed to open-market rents. In the circumstance of an 'open-market' review, the approach to assessing rent is often by reference to 'comparable' market transactions – for example, new open-market lettings. What is vital is that if comparables are considered, then any assessment must be made, not just in terms of factors such as location, but also lease terms and importantly, specification. In the limited-service sector, tenants can be provided with a fully-fitted/turnkey property, a first fix/warm shell or a developers' shell. Whatever the specification, the comparables have to be comparable – or as our US cousins would say – you have to compare 'apples with apples'.

In the full-services sector, comparables are still important. However, the initial approach to rent reviews is often by reference to the trading potential of the hotel. This methodology considers the level of trade a reasonably competent operator (as opposed to the actual operator) could achieve. A 'divisible balance' is assessed and part of this (the tenant's bid) is paid to the landlord as rent and part as 'kept' by the tenant as reward for operating the hotel. The nuances of a profits approach to assessing the open-market rental value are important to understand.

Therefore, it is imperative that both landlord and tenant are represented by experienced surveyors who understand the hotel sector and the intricacies of rent reviews.

Comparable methodology

There are very rarely two hotels in similar locations, with similar facilities, that would be trading in a similar enough manner to enable the rent offered in one hotel to be comparable enough to do more than be used to set a general tone for the letting negotiations.

Comparable evidence tends to be used as the sole method of valuation in cases where trading is not the motivating factor behind the letting, for example, in 'lifestyle' properties. It is also used to check the results when the profits method has been used, to ensure the end result seems reasonable compared with the 'market tone'.

It is usual when looking at comparable rental transactions to adjust these as appropriate to reflect a number of relevant factors including the prevalent trading conditions in the locality, the relative size and facilities of the hotel, as well as differences in dates, lease terms and underlying demand in the market.

Example: Hotel Warne

Hotel Warne is one of a number of terraced guest-houses located along the seafront, all of which cater almost exclusively to the local domestic bed and breakfast leisure market. Each property is arranged on lower ground, ground and four upper floors and has between eighteen and twenty guestrooms, depending on the internal layout. The landlord for all the hotels is the same, and the lease terms are identical.

The new lettings and rent reviews in Table 11.1 have happened in the last three years.

There is a definitive pecking order for evidence, with new lettings holding more weight in terms of reliance as comparable evidence than rent reviews, which are more artificially arrived at.

Rental evidence from the analysis of comparable lettings will usually be the most highly prized evidence. It is important, however, for valuers to attach the relevant weight and importance to all the comparable evidence. Greatest weight will be given to transactions relating to properties as similar as possible to the subject property, and where the date of the transaction is as close as possible to the valuation date for the rent review in dispute.

Valuers may also be aware of transactions occurring after the date of valuation (in the event of a rent review). Such transactions are not primary (or indeed genuinely relevant) evidence, since they could not have been known to the parties on the review date. They can, however, be useful to support the valuer's opinion as to movement in the market place.

The quality of the evidence derived from different sources will depend upon the circumstances of each case. The descending order of weight is shown in Figure 11.1

Table 11.1 Comparable evidence

Hotel	*Beds*	*Rent*	*Rent/ bedroom*	*Date*	*Type*
Hotel Haydn	18	109,800	6,100	2 months ago	New letting
Hotel Waugh	18	108,000	6,000	1 year ago	Rent review
Hotel McGrath	20	122,000	6,100	1 year ago	New letting
Hotel Ponting	20	129,000	6,450	2 years ago	Rent review

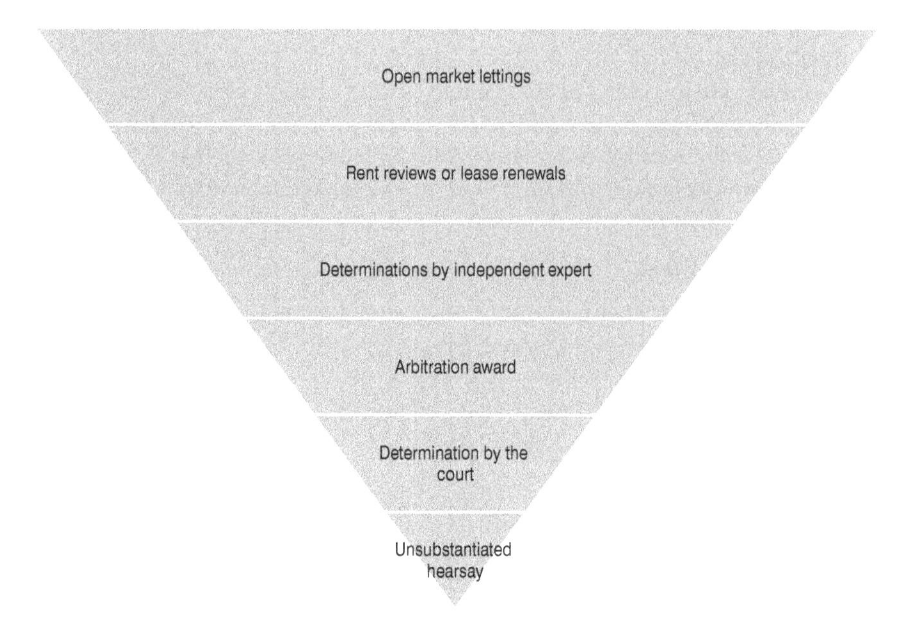

Figure 11.1 Weight of evidence for comparable evidence

Open market lettings

An open market letting is the most important evidence when assessing market value. If the property has been adequately exposed to the market, and both parties have been well advised then the terms agreed should represent market rental value of that specific property.

Agreements between valuers at arm's length upon lease renewal or rent review

An agreement between valuers (for example, rent reviews or lease renewals) will usually reflect most of the evidence in the market at the relevant date, and it can usually be assumed that the negotiations, if both parties had equal negotiating positions, are likely to have reached a fair conclusion.

It is important, however, to be aware of and judge the market knowledge of the parties involved. It can happen that an experienced valuer can 'pull the wool over the eyes' of someone less experienced. The same can happen when one party is unrepresented. The position in respect of security of tenure is also important when assessing such agreements. A tenant at the expiry of a contracted-out lease is in a potentially weak negotiating position and may be willing to pay a rent in excess of the normal open-market level in order to avoid the cost of moving to new premises. As with all comparable evidence, it is essential for valuers to make enquiries and to be aware of the background factors behind the agreement.

Determination by an independent expert

A determination by an independent expert is reliant upon the quality of that such expert. However, the likelihood is that representations will have been received from both parties. Consideration of all the comparables put forward by the parties, as well as any comparables known to the expert, will have been undertaken. Such a determination therefore ranks very closely with an agreement between valuers in the weight that ought to be accorded to it.

An arbitrator's award

The value of an arbitrator's award as evidence, generally, depends entirely upon the quality of the evidence put to him during the case, the manner in which it is expressed and the manner in which the parties were represented. Often an arbitrator may determine a figure in the light of the evidence which is not his personal opinion of the true market value. The weight that should be attached to the award will therefore vary, depending upon the quality of evidence, the arguments put before the arbitrator and the quality and market knowledge of the arbitrator and the surveyors acting on behalf of the parties.

Determination by the court under Part II of the Landlord and Tenant Act 1954

An arbitrator's award will rank ahead of a judge's determination of the rent payable under a new tenancy granted pursuant to Part II of the Landlord and Tenant Act 1954, because the arbitrator possesses the requisite expertise in the field with which to evaluate the parties' evidence and submissions, where a judge rarely will.

Unsubstantiated hearsay

Unsubstantiated hearsay can be beneficial to review 'market perception' at a relevant date, but without evidence that the facts are correct, very little weight can be attached to such evidence.

Taking the relevant weight of the transaction types into account, the Hotel Haydn is clearly the most useful single comparable, followed by the Hotel McGrath, despite it being nearly a year out of date.

As such, it is clear that all the evidence points to a rental value for the building of £62,000, or $6,100 per bedroom, and so the valuer in this valuation is likely to assess the rental value for the property at $122,000 per annum.

Example: Hotel Roberts

In this example, four family-run 'lifestyle hotels' have recently been leased and have all been considered to some extent to be relatively comparable to Hotel Roberts, the subject property.

The Hotel Roberts is a four-star property located in the city centre with excellent visibility; it has 100 bedrooms, extensive conferencing facilities, good leisure facilities, good car parking and is well regarded in the marketplace. It is approximately ten years old and trades mainly to corporate clientele, with a good selection of individual leisure guests and wedding business at weekends. The lease has an unexpired term of fifteen years, with five-yearly rent reviews on FRI (full repairing and insuring) terms to a shell and core specification.

The four comparables can be summarised as follows:

1. The Hotel Croft is not as well located, close to the station just out of the city centre in a slightly rundown part of the city. Surrounding the hotel are a number of cheaper hotels and hostels providing accommodation for more price-conscious guests. The hotel caters to mid-market clientele, with lower-rated corporate customers and individual leisure guests. It is about eighty years old, quite tired and has sixty bedrooms, a small bar and breakfast area and no car parking facilities. It was let on a fifteen-year lease, shell and core basis, on FRI terms and the deal was completed two years ago. The rent equates to $17,000 per bedroom and is subject to review every five years.
2. The Hotel Holding is in a similar location to the Hotel Rabat close to the centre of the city and is less than one-year old and provides top-end four-star accommodation. It has 120 large bedrooms, excellent leisure facilities, two small meeting rooms, a fashionable restaurant (with a celebrity chef) and a top rated spa. Parking is limited. It was let four months ago on a sale and leaseback on a fully fitted basis, paying 21 per cent of turnover, of which $18,000 per bedroom was guaranteed. The guaranteed part of the rent was annually reviewed to RPI. The total rent is anticipated to be in the region of $22,500 per bedroom.
3. The Hotel Garner is located in the tourist section of the city and caters mainly to bus and coach tours. It has 400 bedrooms, large restaurant and bar areas and is forty years old. It was let fifteen years ago on a thirty-five-year lease on a fixed rent, with the last review one year ago showing a rent of $14,750 per bedroom, on FRI terms, on a shell and core specification, with reviews every seven years.
4. The Hotel Marshall is a budget hotel located on the outskirts of the city which has just been opened with 100 bedrooms. It has minimal public areas, well-designed bedrooms and good parking. The property was let on a shell and core basis two months ago, on FRI terms, with annual RPI-linked rent reviews and equates to a rent of $14,200 per bedroom. The hotel is part of a larger development that took two years to construct, and is now 80 per cent open. The heads of terms for the lease were signed and agreed two years ago.

The comparable evidence in Table 11.2 suggests that the rental value of the property will be somewhere between $14,200–$22,500 per bedroom or $1,420,000–$2,250,000 per annum, depending on how the evidence is analysed.

Table 11.2 Comparison of hotels

Property	*Date*	*Rooms*	*Reviews*	*Rent/room*
Hotel Croft	2 years ago	60	5-yearly reviews	$17,000
Hotel Holding	4 months ago	120	Annual review, part to RPI, fully fitted	$18,000–22,500
Hotel Garner	1 year ago	400	Every 7 years, shell and core specification	$14,500
Hotel Marshall	2 months ago	100	Annual review to RPI	$14,200
Hotel Roberts (Subject)	Now	100	5-yearly rent reviews, shell and core specification	?

It is essential that the valuer has sufficient experience and knowledge of the comparable transactions to be able to accurately analyse the relevant comparable evidence.

In this particular example, the valuer was able to make the following adjustments.

Hotel Croft

- + 2.5 per cent for the size differential
- +2.5 per cent for the location
- +2 per cent for the age difference
- +5 per cent for the time difference between this letting and the valuation date
- 0 per cent for the difference in facilities at the hotel
- 0 per cent for the difference in the overall quality of the property
- 0 per cent difference for the nature of the rent review clause and the building specification
- + 12 per cent total adjustment.

Hotel Holding

- 0 per cent for the size differential
- 0 per cent for the location
- –10 per cent for the age difference
- 0 per cent for the time difference between this letting and the valuation date
- –10 per cent for the difference in facilities at the hotel
- –10 per cent for the difference in the overall quality of the property
- –10 per cent difference for the nature of the rent review clause and the building specification
- –40 per cent total adjustment.

Hotel Garner

- –7.5 per cent for the size differential
- +10 per cent for the location

- +2 per cent for the age difference
- +2.5 per cent for the time difference between this letting and the valuation date
- +2 per cent for the difference in facilities at the hotel
- +2 per cent for the difference in the overall quality of the property
- +2 per cent difference for the nature of the rent review clause and the building specification
- + 13 per cent total adjustment.

Hotel Marshall

- 0 per cent for the size differential
- +25 per cent for the location
- –10 per cent for the age difference
- +10 per cent for the time difference between the date the rent was agreed and the valuation date
- +5 per cent for the difference in facilities at the hotel
- +10 per cent for the difference in the overall quality of the property
- –10 per cent difference for the nature of the rent review clause and the building specification
- + 35 per cent total adjustment.

Summary

On this basis, the value of the hotel is likely to lie within a range of $13,500–$19,040 per bedroom ($1,350,000–$1,904,000) (Table 11.3).

The valuer will then be expected to use their judgement to determine where within that range this particular property would sit. In this instance it has been determined that the value would be $16,200 per bedroom, equating to $1,620,000 per annum rent.

It should be noted here that in Table 11.3, adjustments are specific to this example; all adjustments used in practice will be determined as much as possible by valuation evidence and will otherwise be determined by the personal experience of the individual valuer.

It should also be noted, however, that almost all comparables will need to be adjusted to look at the underlying potential profitability of the hotel. As such, the key method of assessing hotel rents is still the profit method of valuation.

Table 11.3 Summary of adjusted valuations

	Rent/bedroom	*Adjustment*	*Adjusted rent/bedroom*
Hotel Croft	$17,000	+ 12%	$19,040
Hotel Holding	$22,500	– 40%	$13,500
Hotel Garner	$14,750	+13%	$16,667
Hotel Marshall	$14,200	+ 30%	$18,460

Profits method

However most 'arm's length' rental levels are agreed based almost exclusively on the potential profitability of the specific hotel. Almost without fail, any operator looking to lease a hotel will review the trading potential of a particular unit before determining the level of rent that can be afforded. Paying too much rent would be detrimental to the long-term future of the operational business, whilst too low a rent would adversely impact on the investor's income stream.

The usual methodology for assessing the rental value of a property is to determine the 'divisible balance' and then apportion that amount between the landlord and the tenant.

Calculating the divisible balance involves working out the EBITDA for the property and then deducting a number of additional items including an annualised sum for the tenant's fixtures and fittings and a sum for working capital.

The apportionment of the divisible balance reflects 'the tenant's bid'. If the market is performing strongly and the property is desirable, the tenant will be prepared to offer more of the divisible balance to secure the property. Conversely, if the market is weak and there is less demand for the property, the landlord will be prepared to accept a lower proportion of this profit for the property.

It is then normal practice to review this 'rental bid' as a proportion of turnover, as a proportion of EBITDA and on a rent per bedroom basis in light of other market transactions.

Unfortunately, it is not always possible for the valuer to look through the trading accounts of the hotel when assessing the rental value of the property; ideally, the valuer would have access to such accounts but as few traditional leases provide for such information to be made available, the assumed trading projections are normally down to the experience of the valuer.

As such, it would be important for the valuer to try and ascertain the stabilised trading as described earlier in financial and valuation due diligence chapters in Part II. It may not be possible for the valuer to interview the general manager of the hotel as sometimes the information provided may not be as full and frank as the valuer would hope for. Once again, experience in the local market is essential to help overcome such shortcomings.

Example: Hotel Muralitharan

The valuer will need to take into account exactly what it is they are trying to value. The lease will detail the actual demise of the property (for example, does it include the FF&E) and will outline the terms on which it will be valued.

The performance of the hotel is assessed by the valuer in today's values, based upon the demised premises; for example, if the hotel was let five years ago and was in good condition at the time of the letting, then the valuer needs to assess how that hotel would trade in today's market.

The process is very similar to that of a capital valuation all the way through to the IBFC line; the occupancy rate and ADR are assessed to calculate rooms'

Table 11.4 Hotel Muralitharan (valued in JPY)

Valuation date	1 January 2015		
Number of bedrooms		60	
Occupancy rate		72.50%	
ADR		2,000,000	
Total rooms revenue		31,755,000	
Rooms rev/total revenue		65.00%	
Total revenue		48,853,846	
Direct room costs	33.00%	10,479,150	
Other revenue costs	38.00%	6,412,067	
Unallocated costs	16.25%	7,938,750	
IBFC		24,023,879	49.18%
Fixed costs			
Property taxes		1,545,000	3.16%
Management fee		1,465,615	3.00%
Insurance		488,538	1.00%
Total		3,499,154	
Tenants FF&E			
	Rooms	60	
	@	750,000	
Total		45,000,000	4,500,000
Working capital	3 weeks' revenue	2,818,491	
EBITDA			20,524,725
Divisible balance			13,206,234
% as rent	65.00%		8,584,052
Rental value			8,584,052
Say			8,585,000
Price per bedroom			143,083
% of turnover			17.57%
% of EBITDA			41.83%

revenue, which is then worked into total revenue. Costs are then allocated across each department and in unallocated costs until the IBFC has been assessed.

It is then market practice to deduct fixed costs (except rent) including rates, management fees (where it is appropriate), FF&E reserve (where appropriate) and insurance costs.

Then the adjustments that are specific to rent review valuations are made. Adjustments are made to the profit margins to take into account other items, for example, working capital and the cost of furnishing the property (assuming it has been let on a shell and core basis with no FF&E included).

The adjustment for fitting out should be made as a capital amount (probably as the cost per bedroom to provide a suitable level of FF&E for the hotel type) and then annualising it over a suitable period of time (ten years seems to have come into favour as a conventional period of time but there is no reason why this should not be altered if a more appropriate period can be shown).

It is deducted because the hotel will not be able to trade without FF&E, and it is presumed that the hypothetical operator will need to invest the money to furnish the property and as such, the cost needs to be taken into account.

The working capital is also a cost of operation, and convention dictates that three weeks' revenue is deducted from the profit to make an appropriate allowance for this expense.

This works down to a divisible balance (which is lower than the EBITDA of the property because of the additional deductions). The divisible balance is the amount of profit available at the hotel that needs to be divided between the landlord and the tenant (between rent and retained profit).

The decision on how to assess the proportion of divisible balance that the hypothetical tenant will be prepared to pay as rent to the landlord is quite difficult and will be based upon a number of things that will require the valuer to have a good knowledge of the local market. Prevailing economic conditions, demand for hotels from operators, hotel trading conditions (both the supply and demand inherent in the market place), as well as the quality of the asset and its suitability and positioning within the local hotel market are all items that will have an effect on the amount of the divisible balance that will be offered to the landlord. Needless to say, comparable evidence here is very important, but needs to be analysed and used on a consistent basis.

Further adjustments will then be made for specific lease terms that are unusual and for peculiarities in the rent review period.

It is then usual for the valuer to look at the result to assess whether it seems in line with the general market. There are three good tests for this: rent as a per cent of turnover, rent as a per cent of EBITDA and to a lesser extent rent per bedroom. This 'stand back and review' of the results is a vital part of the valuation process.

Sometimes of course it is less straightforward, as shown in the next example.

Example: Hotel Stokes

In this example, the hotel was let thirty-five years ago on an eighty-four-year lease and since that time, substantial tenant's improvements have been undertaken to make the bedrooms attractive to modern hotel guests, with air conditioning and en suite facilities having been introduced. However, the rent review clause specifically disregards any tenant's improvements from being included when assessing the market rent.

At the start of the lease, the hotel had forty-two non-en suite rooms and twenty en suite rooms; the present bedroom configuration is forty-eight en suite bedrooms.

In this instance, there are two approaches the valuer will have to adopt to accurately assess the rental value of the property. A calculation will need to be carried out on the property as it originally was (with sixty-two rooms), assuming the valuer believes there are still potential tenants who would be prepared to operate a hotel in that condition (Table 11.5).

The second approach will be to assume the improvement works have been carried out at the expense of the 'hypothetical tenant' and to capitalise this expenditure (Table 11.6).

In this instance, both methods provide the same value. In the event of significant differences, the valuer will need to determine which method would most accurately represent the prevalent market approach.

The 'profits method' of valuation is used at almost every rent review (if the review is to 'market rent') and for all new lettings as it is the most effective way of determining the rental level of a property, where the reason behind a potential tenant's occupation is to operate a financially successful business.

However, if the lease determines a different method of valuation (for example, it is assessed to 4 per cent of the capital value or assessed to 80 per cent of office rental on the adjacent building) then the specified methodology will need to be adhered to.

It is not always used when assessing lifestyle properties, as the motivation behind occupation is nothing to do with running a successful business and may therefore attract bids that are calculated without reference to profitability.

The impact of lease terms on rental values

The terms and covenants contained in the lease may directly impact on the profitability of the hotel, or they may impact on the desirability of the property from an operational perspective. As such, the lease terms will have a key impact on the assessment of market rent.

The exact wording of the lease, and the terms contained, will impact on the rental value of the property. One of the most important terms will be the rent review clause and the key assumptions contained within it, including the following.

The assumed term

Is the assumed term longer or shorter than the market average, and would this impact on the rent that a hypothetical tenant would be prepared to accept?

When the length of 'notional term' in the rent review clause is different to terms commonly accepted in the market place, this could have an effect on the rental value at the review date. An assumed term at rent review of forty years without a break, when the norm in the open marketplace is for a term closer to twenty years, could potentially lead to an argument for a discount in the revised rent.

Table 11.5 Hotel Stokes – first approach (valued in £s)

Valuation date	25 December 2005		
Number of en suite bedrooms		42	
Occupancy rate		77.00%	
ADR		115.00	
Number of non-en suite bedrooms		20	
Occupancy rate		77.00%	
ADR		95.00	
Total rooms revenue		1,891,467	
Rooms revenue/total revenue		85.00%	
Total revenue		2,225,255	
Direct room costs	27.00%	510,696	
Other revenue costs	60.00%	200,273	
Unallocated costs	23.00%	511,809	
IBFC		1,002,477	45.05%
Fixed costs			
Rates		86,117	3.87%
Management fee		66,758	3.00%
Insurance		22,253	1.00%
Total		175,128	
Tenants FF&E			
	Rooms	42	
	@	12,000	
Total		504,000	50,400
	Rooms	20	
	@	6,500	
		130,000	13,000
Working capital	3 weeks' revenue	128,380	
Divisible balance			635,570
% as rent	55.00%		349,563
Rental value			349,563
Say			350,000
Price per bedroom			5,645
% of turnover			15.73%
% of EBITDA			47.40%

Table 11.6 Hotel Stokes – second approach (valued in £s)

Valuation date	25 December 2005		
Number of en suite bedrooms		62	
Occupancy rate		77.00%	
ADR		115.00	
Number of non-en suite bedrooms		20	
Occupancy rate		77.00%	
ADR		95.00	
Total rooms revenue		2,003,887	
Rooms revenue/total revenue		85.00%	
Total revenue		2,357,514	
Direct room costs	27.00%	541,049	
Other revenue costs	60.00%	212,176	
Unallocated costs	23.00%	542,228	
IBFC		1,062,060	45.05%
Fixed costs			
Rates		91,236	3.87%
Management fee		70,725	3.00%
Insurance		23,575	1.00%
Total		185,536	
Tenants FF&E			
	Rooms	62	
	@	12,000	
Total		744,000	74,400
	Other improvements	1	
	@	350,000	
		350,000	35,000
Working capital	3 weeks' revenue	136,010	
Divisible balance			631,113
% as rent	55.00%		347,112
Rental value			347,112
Say			350,000
Price per bedroom			5,645
% of turnover			14.85%
% of EBITDA			44.74%

Figure 11.2 Hotel leases on ground rent

Figure 11.3 Site leased on turnover rent

Conversely, however, if the assumed term is too short to allow a tenant a sufficient period to write off fitting out expenses (in the case of a non-fully fitted property) and marketing set-up expenses, the rental value could be reduced.

When determining whether the assured term is too long or too short, the size of the hotel can sometimes be a relevant factor. Tenants leasing a hotel with over 100 bedrooms will be unlikely to accept a shorter term unless the rent is lower, whereas a tenant may well be prepared to let a smaller hotel with twenty bedrooms on a short-term basis.

In the UK the effects of unusually shortened terms were to some extent mitigated by Pivot Properties Ltd v Secretary of State for the Environment. This case established that the valuer must take into account the possibility of renewal under Part II of the Landlord and Tenant Act 1954.

However, it is the duty of the valuer to reflect the open market; if the market is likely to worry about any of the section 31 grounds for possession being enforced, particularly the use by the owner or for redevelopment, then the rent will be discounted by the shorter term.

The section 31 grounds are as follows:

> (a) where under the current tenancy the tenant has any obligations as respects the repair and maintenance of the holding, that the tenant ought not to be granted a new tenancy in view of the state of repair of the holding, being a state resulting from the tenant's failure to comply with the said obligations;
>
> (b) that the tenant ought not to be granted a new tenancy in view of his persistent delay in paying rent which has become due;
>
> (c) that the tenant ought not to be granted a new tenancy in view of other substantial breaches by him of his obligations under the current tenancy, or for any other reason connected with the tenant's use or management of the holding;
>
> (d) that the landlord has offered and is willing to provide or secure the provision of alternative accommodation for the tenant, that the terms on which the alternative accommodation is available are reasonable having regard to the terms of the current tenancy and to all other relevant circumstances, and that the accommodation and the time at which it will be available are suitable for the tenant's requirements (including the requirement to preserve goodwill) having regard to the nature and class of his business and to the situation and extent of, and facilities afforded by, the holding;
>
> (e) where the current tenancy was created by the sub-letting of part only of the property comprised in a superior tenancy and the landlord is the owner of an interest in reversion expectant on the termination of that superior tenancy, that the aggregate of the rents reasonably obtainable on separate lettings of the holding and the remainder of that property would be substantially less than the rent reasonably obtainable on a letting of that property as a whole, that on the termination of the current tenancy the landlord requires possession of the holding for the purpose of letting or

otherwise disposing of the said property as a whole, and that in view thereof the tenant ought not to be granted a new tenancy;

(f) that on the termination of the current tenancy the landlord intends to demolish or reconstruct the premises comprised in the holding or a substantial part of those premises or to carry out substantial work of construction on the holding or part thereof and that he could not reasonably do so without obtaining possession of the holding;

(g) subject as hereinafter provided, that on the termination of the current tenancy the landlord intends to occupy the holding for the purposes, or partly for the purposes, of a business to be carried on by him therein, or as his residence.

Of particular interest to valuers is the repossession for their own use of the property by landlords. It should be noted that it is considered to be taking in a property for one's own use if the intention is to let a manager operate the property on a management contract.

Assumptions as to repair

There are a number of considerations that must be taken into account in relation to the assumed condition of the premises for the purposes of review.

If the hotel is let on full repairing terms, then the rent review clause usually dictates assumption of compliance by tenant with covenants. Where the review clause requires the valuer to assume that all the tenant's covenants have been complied with, no valuation problems should arise. Where the lessee is responsible for all repairs, the property will be valued as if in good repair.

In rare cases, there may be an issue as to whether defects existing at a review date are matters of repair covered by the repairing covenant or would require replacement on a scale which would put them outside the scope of the repairing obligation, as shown in Ravenseft Properties Ltd vs Davstone (Holdings) Ltd (1980).

The existence of a defect for which the tenant is not responsible could have very serious valuation consequences – as shown in Camden LBC v Civil Aviation Authority and Langford (1980) – a rating case where defects were held to make the property virtually unlettable.

If the landlord has covenanted to repair and the property is in disrepair, it is a matter of judgement as to whether the right for the tenant to enforce the landlord's covenant would wholly or partially offset any diminution in value attributable to the disrepair. In practice, any material disrepair is likely to adversely affect the rental value.

If the rent review clause stipulates that it shall be assumed that both landlord and tenant have complied with their repairing covenants, the valuer must value as if the property is in good repair, even if this produces an injustice to the tenant, where the property is in disrepair through failure by the landlord to perform his covenant.

Alterations and improvements

It is usual for the lease to disregard tenant's improvements. This can however have quite a dramatic effect on the value, where improvements rather than normal upgrades have been made.

It is essential that the valuer knows exactly what they are valuing when trying to assess the rental valuation of a property. Questions such as, 'Have any tenant's improvements been carried out that are excluded from the valuation? What is the specification of the building to be valued or does the demise of the leases include the furniture?' should be foremost in the valuer's mind.

Rent review patterns

If the timing of the rent reviews provided for in the lease is atypical of the wider market, it can impact on the market rental value of the property. If the normal rent review pattern was a review every five years' it is possible that a ten-year review pattern would lead to higher market rent for that property, as the delay in reviews is valuable to the tenant and would potentially lead to a higher 'tenant's bid'.

Other methodologies for assessing rental valuation

There is another method of assessing rents that is sometimes used.

Rental levels as a per cent of total investment

In certain circumstances, the rental level of a hotel may be initially based on building costs where an operator will offer a certain guaranteed return based on total investment costs. These initial offers are always reviewed to ensure the rent is 'affordable' by the tenant.

Example: Hotel Gavaskar

A developer has secured a site for a new hotel and is building a 210-bedroom hotel which will be let on completion to a highly respected operating company. The land is costing £2,000,000, the construction of the hotel to shell and core specification will cost an additional £19,000,000 and with fees, finance costs, etc. the overall cost of the development is expected to be £23,800,000. The operator has offered to provide a rent based upon a 6.75 per cent gross return of the total investment cost, to be annually reviewed to RPI (or £1,606,500 per annum).

Rent = 6.75 per cent × £23,800,000 = £1,606,500

Note that the agreed return will generally be higher than the market price for an operational investment with the same operator to reflect the added development risk.

Table 11.7 Hotel Gavaskar

Cost of land	£2,000,000
Cost of building	£19,000,000
Other costs	£1,800,000
Total costs	£23,800,000
Gross return required	6.75%

In development scenarios like this, the developer will normally try and ascertain rental bids from similar quality operators (in terms of covenant strength) based on a percentage of the total cost of developing the hotel.

This method is mainly used when considering the level of rent an operator should offer prior to the development being completed. It is also used as a check method for market rents for other properties.

12 Site values and how they are determined

Introduction

This chapter will explain how site values are determined, how hoteliers and developers decide where new hotels or resorts should be built, and indeed how it can be determined whether the development is feasible.

The theory is simple; new hotels and resorts will only be considered feasible by developers if they are able to be sold for more money than they cost to construct. Hoteliers may sometimes have different criteria as an operating presence in certain markets might be more valuable to the hotel group as a whole than individual profits from one property. However, as a rule of thumb, most developers would not tend to develop the property if the end value is less than the development costs.

Basic valuation methodology

The basic methodology behind a site valuation is once again twofold. The valuer will look at any comparable transactions to refer to a price for the land first and foremost to see the range in which it is likely that the site value will sit.

Unfortunately, because of the specialist nature of a hotel's trading profile, and because there are rarely sufficient comparable site transactions to refer to, it is usually necessary for a valuer to resort to undertaking a residual valuation.

Various concerns over the adequacy of the residual method of valuation have been raised by a number of practitioners and courts. The large number of variables used in arriving at the end value, and the sensitivity of the end value to even minor changes in these variables, has meant it is even more important than usual 'to stand back and review the result'.

As such, it is always important that the valuer highlight in their report these potential inadequacies in the methodology and draw the client's attention to the uncertainty in the end value.

That said, the market does still carry out residual valuations in the majority of cases to determine the price that can be paid for a site. As such, it is probably the best method for arriving at the value of the site if the valuer is aware of the variables used by the majority of the active purchasers in the market place at the time of the valuation.

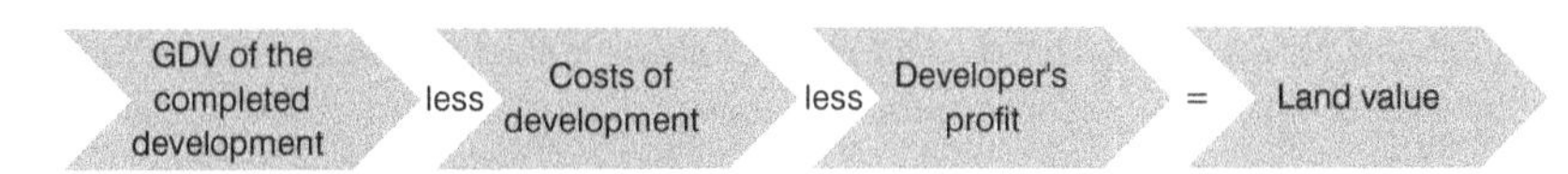

Figure 12.1 Residual valuation process

In simple terms, the residual method of valuation can be summarised as in Figure 12.1.

This is illustrated further in the next example.

Example: Hotel Flower

The gross development value (GDV) for the hotel is calculated at ₦20 billion. The cost of the development (including finance) is ₦14 billion. The developer's profit is assessed to be 25 per cent of development costs which is ₦3.5 billion. Therefore, the gross land value is ₦2.5 billion (Figure 12.2).

The GDV is the estimated sale price of the completed development, and will be calculated taking into account the trading profile of the hotel throughout the building up period and then multiplied by the appropriate multiple.

The costs of development will include, where relevant, the following:

- demolition works
- site remediation works
- construction of the hotel
- ground works
- fees for professional advisors, including architects, quantity surveyors, structural engineers, project managers, etc.
- finance costs
- transaction costs
- FF&E
- pre-opening expenses.

The developer's profit will be the return that the developer will require to carry out the project and is usually expressed as a per cent of the total development costs (including the land), although it is sometimes referred to as a per cent of the GDV.

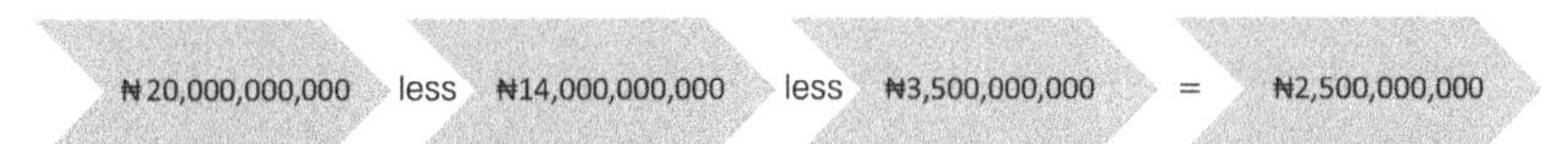

Figure 12.2 Gross land value calculation

The land value then needs to incorporate the cost of purchase and the cost of finance to arrive at the net land value. This value is then usually calculated on a per bedroom basis to see if it is in line with usual market values.

Calculating the GDV

It is essential that the valuer explores the characteristics of the existing site and has an adequate knowledge of each of the development components. The level of detail that is appropriate when assessing development potential varies according to the purpose of the valuation. Indeed, the level of information available for the valuation is usually dependent upon what stage the valuation is carried out. For example, a valuation in advance of planning consent being gained is based on less certain estimates than if the planning has been progressed, or the valuation is at a date where the redevelopment has already commenced.

It is much more difficult for a valuer to determine the trading profile for a new hotel compared to an existing hotel at some point in the future, as there is no trading history to base future earnings on, and the position, style, design, corporate environment and staffing levels are all relative unknowns.

It is usual for a developer, when looking at developing a new hotel, to commission a feasibility study to ensure that the development makes sense. This is to try and ascertain with some degree of certainty how the completed hotel is likely to trade.

The valuer will review the feasibility study, casting an expert eye over the commentary and projections, to see if they tally with their experience of the local hotel market. In some instances, the valuer will have to make adjustments to the projections to tie in with what they feel the market would expect, in such a location, from such a type of hotel.

Inspection and site-specific information

The physical inspection of the site, and the related enquiries, will usually reveal site-specific information that is relevant to the type of development proposed. Such information could potentially include the following:

- *The extent and dimensions of the site.* It is of paramount importance to understand the dimensions and range of the site in order to ascertain frontage, width and depth, as well as gross site areas and net developable areas of the site. These factors are key to the likely profile of all developments.
- *The shape and slope of the site*, ideally in the form of a topographical survey. This is important, as the topography can limit the 'developability' of sites, making similar-sized sites significantly more or less valuable, purely based on this factor alone.
- *The position of the site* in relation to surrounding land uses, including complementary land uses and those that may detract from the proposed use of the developed property. This is obviously of paramount importance to

Beware the seven-star visionary!

James Bradley – Savills

Greece is a country of historic sites, golden beaches, fine food and a wealth of hotels. It also benefits from the euro (for now at least). While relationships with some of its European creditors may be strained, the masses still flock to Greece's sandy shores – tourism contributes around 20 per cent to total GDP. By the beard of Zeus!

Undertaking a hotel development valuation in Greece is usually a pleasure, though certainly comes with its challenges. Not least of which can be the over-enthusiastic developer.

His or her hotel will be the best on the island! It will have the longest infinity pools, the whitest private beach, the classiest restaurants ... probably. If next-door is six-star, this will be seven-star ... probably. There have already been ten offers for the site ... possibly. Ouzo may be deployed to help you see this vision. (The true professional will resist ... probably.)

However, you are still looking at ten acres of dirt.

The lesson for us humble hotel valuers – balance the art of valuation with the science of valuation. Have the imagination to see the vision but the prudence to anchor your valuation firmly in facts and reasoned deductions.

What should we focus on for such a Mediterranean development?

- *Route to market*. What are the guest source markets, how frequent are the flights, what is the guest profile? Can you rely on 'build it and they will come'?
- *Seasonality*. For how many months can you drive occupancy? For example, Crete has a significantly longer season than Corfu. No one drinks piña colada in the rain!
- *Comparables*. The backbone of any valuation. Transactions may be few and opaque, but yields, prices per bedroom and prices per square metre should be sought and analysed.
- *The build*. Are all permits and permissions in place? How easy is it to connect to public roads? Is the development suitable for its location and market?
- *Operations*. Does the team have the experience? What (if any) branding will be used – Ikos, Banyan Tree, the Luxury Collection (Starwood) or Six Senses (to name but a few)? Are the management agreement terms market-based? And how does the hotel's competition trade?

The conclusion? There are challenges in valuing a resort hotel but substantial potential for the right scheme. At the end of the day, the valuer needs to assess whether the euros will flow or will the investor get (golden) fleeced?

the likely success of the new development, with a positive alignment with surrounding land uses enhancing site value.

- *The history of previous land uses*. Previous land uses that may have contaminated the land will detract from the site value, as depending on the type of potential contamination, they may involve significant decontamination costs, or in certain circumstances, preclude certain types of redevelopment altogether.
- *The risk of future flooding*. The likelihood of future flooding is becoming more and more important, as weather patterns have been changing. If flooding is possible, or likely, it may deter certain purchasers from buying the site. Alternatively, it may increase operational costs like insurance, reducing the price that can be paid for the completed property.
- *The size and height of any existing buildings on the site*. This is of course of vital importance if the existing buildings are intended to form part of the final development, as their size, suitability and compatibility with the rest of the development will have a direct impact on the resulting GDV. However, in the event the buildings will not form part of the final development, the size of such properties will still have an impact, this time because of the cost of their removal.
- *The building height and size of adjoining properties*. As a rule of thumb, if the adjacent buildings are all developed to fifteen storeys height, then it is likely you may also be allowed to develop to fifteen storeys height. In the case of confined sites, the height and position of adjoining buildings may impact on your development, whether through rights of light or visibility issues.
- *The efficiency and suitability of existing buildings*, if they are to be retained. In addition to the point above about size and height, the suitability of existing buildings for the proposed use, as well as the level of efficiency in terms of design for the new use, will have an important bearing on the GDV. In addition, efficiency in terms of energy efficiency taking into account new ecological parameters being introduced each year such as the building research establishment environmental assessment method (BREEAM) are all also important.
- *Any matters that may result in excessive abnormal costs* (such as constrained site conditions, and poor or limited access), from development and occupational perspectives. The importance of this is relatively obvious, and all items that could enhance development costs or occupational costs should be reviewed carefully and factored into the development costs or GDV calculation.
- *Any party wall issues or boundary issues*. Boundary issues can be troublesome and relatively expensive to resolve, so the inspection should identify any such issues ahead of the development costs being finalised so any enhanced costs can be factored in.
- *Geotechnical conditions*
- *Evidence of, or potential for, contamination*. The evidence of certain types of contaminants will detract from the site value, as depending on the type of

potential contamination, they may involve significant decontamination costs, or in certain circumstances preclude certain types of redevelopment altogether.

- *The availability and capacity of infrastructure* (such as roads, public transport, mains drainage, water, gas and electricity). If such services are easily accessible and available, it will reduce required infrastructure investment to provide such services on the site, thereby reducing the overall development costs. If such services are unavailable, it will either require enhanced investment to provide such services, or alternatives will be required which might detract from the final value of the development.
- *Evidence of other head (or occupational) interests in the property*, whether actual or implied by law. The inspection can aid the legal due diligence of a site. Any issues suggesting title might be limited, or there might potentially be interests that would limit the use of the site (whether outstanding legal claims, head leasehold interests of restrictive covenants), would have a negative impact on the value of a site.
- *The physical evidence of the existence of rights of way, easements, encumbrances, overhead power lines, open water courses, mineral workings, tunnels, filling, tipping, etc.* Any encumbrances on free and clear use of a site could potentially have an impact on the overall use of the site, and therefore its GDV.
- *The details of easements, restrictive covenants, rights of way, rights to light, drainage or support, registered charges, etc.*
- *The presence of archaeological features.* These may be evident, or there may be a high probability of their presence due to the site location, for instance, close to city centres. The presence of archaeological remains on a site can significantly delay the development schedule for a project, enhance development costs or potentially halt the entire development.
- *The evidence of waste management obligations* and whether those obligations have been fulfilled. If the obligations have not been fulfilled, the liability to do so may rest with the new site owner, adding to development costs.

In the event that a feasibility study has been commissioned for the site, the valuer will specifically look at what has been proposed in terms of development within that study, to see if, in their opinion, such a scheme is feasible, given the profile of the site. If such a scheme is not deemed feasible, the valuer will amend the details of the feasibility to fit in with their assessment of the potential for the site.

Although the valuation is required of the actual site, there may be a possibility of increasing the development potential by acquisition of, or merger with, adjacent land. Conversely, it may be necessary to acquire adjacent land, or rights over adjacent land, before the proposed development could actually take place.

Planning matters

The existence of an existing planning permission on the site is key to assessing the likely development potential of the site. If for example, outline permission is in place for a 200-bedroom full service resort, then it is reasonable to assume that such a scheme could be developed. Of course, if investigations show that such a scheme could not practically be developed, this will need to be reflected in the assessment of the likely scheme proposed when calculating the GDV.

If a feasibility study has been commissioned, it will usually have reviewed the best use of the site, and may have concluded that a lower density development than is available through the planning permission would be optimal.

The following matters may need investigation:

- *The local regulatory framework in place governing the site.* Typically, if a local framework exists, it will provide outlines for what the site might be used for, density of developments that may be permitted, as well as requirements typically required for certain types of development. So, for example, a certain planning authority might require one parking space per hotel bedroom, whilst another may need one space per ten bedrooms. The cost implications of such policy framework can be important to development viability.
- *The existence of a current planning permission.* This may be outline consent or full consent, and may include conditions or reserved matters. This is fundamentally important, as a site with planning consent has mitigated 'planning risk' from the development parameters, to some extent. The nature of the consent, the scope of the allowed development, as well as overall quantum, are all vital factors in assessing the site value.
- *Where the planning permission is time limited,* it is necessary to establish if it is still valid and, if close to expiry, if a similar permission would be granted again. It is important to consider that timeframes for planning consents expire, and when they have become extant, there is not always the opportunity to gain the same consent again with a new application.
- *Any historic planning consents that were never activated.* In the event that there is no existing consent in place, or that a new development is being considered outside of the scope of any existing consents for development, historic consents may sometimes indicate where possible development constraints may exist. If a consent for a 200-bedroom hotel and resort was granted five years previously but has now become extant, and if nothing significant has changed in the time since it was granted, then it is reasonable to assume that the reasons for granting the first consent may still be in place, so there would be a possibility of the same consent being granted at the current time.
- *Any regulations that specify the extent to which development of the site might be permissible without the need for a planning application or consent.* The possibility to redevelop without the need for planning consent effectively, reducing

Figure 12.3 Hotel site in Spain

the planning risk factor of the development, thereby possibly enhancing the overall likelihood of development and even the site value.

- *The permitted use of existing buildings* (if to be retained), or the possibility of identifying an established use. Ensuring the required use for the proposed development can be achieved legally is of paramount importance to the value of such a development. If an established use can be demonstrated, or even a use swap can be accomplished, it can reduce the planning risk associated with the site development.
- *Legally binding agreements* that have been, or are to be, documented, in order to secure the granting of planning permission. In the event that planning consent has not actually been granted, proof that it will be forthcoming may help reduce the planning risk associated with the site.
- *Any special controls that may apply to the site or buildings included* (for example, conservation area designation, green belt, tree preservation orders, listed buildings, etc.). Any such restrictions can have either GDV implications (where part of the site cannot be redeveloped) or cost implications for the development. In certain circumstances, such controls can actually enhance value as completed developments inside a conservation area, or area of outstanding natural beauty, may command a premium on other, less well located developments.
- *Any requirements to protect or enhance environmentally sensitive features* such as SSSIs (sites of special scientific interest) or water courses, and to comply with the relevant environmental protection legislation. Any such restrictions are likely to have an impact on the potential GDV of the site.

Figure 12.4 Development site in Sri Lanka

- *Any requirements for view corridors, sight lines or buffer zones.* Once again, restrictions on the overall 'developability' of the site are likely to have an impact on the overall GDV.

There are times when the current planning permission is not considered to be the optimum consent for which there is a reasonable prospect, having regard to the applicable planning regime. In such cases, it may be necessary to form a view as to what permission is likely to be obtained and the associated planning agreements that would be required to obtain that consent. This includes consideration of published planning policies, recognising that they heavily influence future additions to the supply of particular types of building.

An accurate assessment of the form and extent of physical development that can be accommodated on the site is essential having regard to the site characteristics, the characteristics of the surrounding area and the likelihood of obtaining permission.

Calculating the cost of development

Assessing the cost of development will include a careful analysis of the timetable for the proposed development, along with a detailed review of each part of the construction programme, to ensure the overall cost of development is adequately reflected.

If detailed cost estimates have been produced by a quantity surveyor, or a design and build contract is in place, these can be extremely helpful in assessing likely building costs. However, it must be remembered that the costs of development are those that would be incurred by the average developer for that site.

Any cost savings that are specific to only one developer may not be relevant when assessing the market value of that site. In the same way, if the development costs proposed by the specific developer are higher than typical market costs, then the proposed costs should be reduced to reflect market averages.

An outline programme may be provided but its achievability needs to be assessed. Such a programme could include the following periods.

The pre-construction period

- This would involve site assembly, obtaining vacant possession, negotiations with adjoining owners, extinguishing easements, or removing restrictive covenants, rights of light issues, etc. It would involve negotiating the planning process, agreeing architectural and engineering design and/or solutions, soil investigations, the building contract tender period, etc.
- It would involve negotiating the form, extent and value of the building contract(s), including demolition and any necessary site preparation. It may be appropriate for the valuer to seek advice from an environmental, quantity or, building surveyor, mechanical engineer or architect to assist them with the process.

The principal construction period

- site preparation
- enabling work may include an archaeological dig, demolition, de-contamination or the provision of infrastructure prior to the main works commencing
- construction.

The post-construction period

- handover to the operator
- rectify defects.

Assessing the timeline for the development will help to accurately assess the cost of financing the development. Typically, the land must be initially purchased, which will also require financing. Development costs will be incurred at various rates throughout the programme, requiring financing at different times. Income from the development will usually not occur until it becomes operational.

Usually, the cost of development is not linear; indeed, development costs are more often based on an 'S' curve, with close to 40 per cent of costs spent in the first half of the development and the remainder spent in the last half of the development.

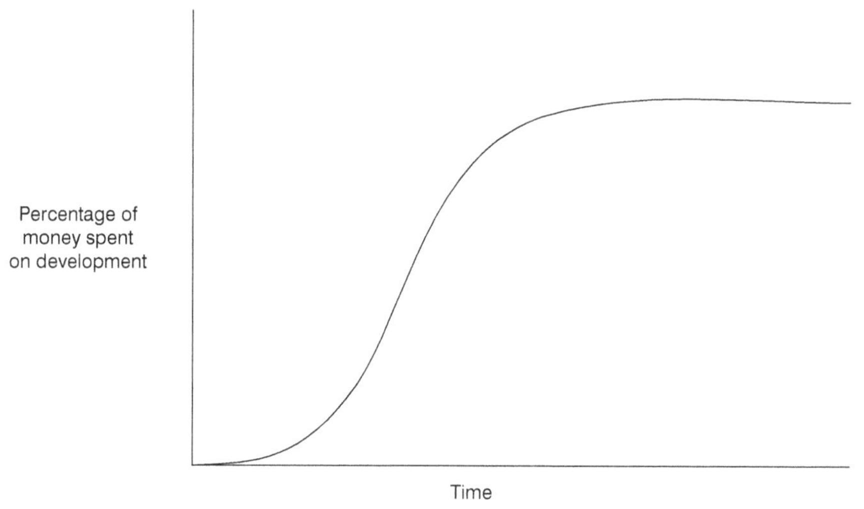

Figure 12.5 Development costs

Demolition works

If the site has existing buildings that are not beneficial to the final development, these will need to be demolished and the rubble will need to be removed, or potentially partially used in the foundations of the new development. The design, type and construction materials used in the existing buildings will have an impact on the demolition costs and must be taken into account.

Site remediation works

Site remediation works can include addressing soil or groundwater contamination, in addition to more complex issues relating to past uses, such as gas works, chemical facilities, pharmaceutical sites and industrial facilities. It can also include removal of sewers, cellars, landfill and any other items that will hinder the suitable development and end use of the site.

Construction of the hotel

This will involve the building costs associated with developing the hotel or resort buildings, including the foundations, the frame of the building, the associated services and the construction finishes of the property.

The developer's preferences for particular design features, building layouts and specifications can be taken into account if that is what would be expected in the wider market place.

Key factors that can impact on construction costs are the availability of the required raw materials and the availability of suitably skilled labour. In certain locations, shortages of one or both can lead to tremendous differences in overall construction costs.

Ground works

This will include the cost of providing the access roads, fencing, as well as the landscaping required to finish off the development and make it serviceable and appealing to customers.

Fees for professional advisors

These fees will include architects, quantity surveyors, structural engineers, project managers, valuers and planning consultants.

FF&E

A hotel or resort cannot become operational until it has been fitted out. It is traditional to separate the cost of the FF&E from the construction costs in a residual valuation, so that an adequate assessment can be made of the allowance for this fit-out. This is quite often an area where cost savings are sometimes sought by the developer, which could have a dramatic impact on the trading potential of the completed development, and therefore the GDV.

Pre-opening expenses

The nature of a hotel is that on the day of completion, it should ideally be full of guests and trading as though it had already been open for some time. As such, most operators will require a pre-opening budget, which will involve training staff ahead of opening, marketing the hotel ahead of opening and ensuring that all the various marketing collateral is in place when the property finally opens its doors to the public.

Finance costs

This will cover the development costs throughout the development period. In addition, a second series of finance costs will usually be required to cover the land acquisition cost, as this is typically required for a longer period, albeit with greater security for the lender, and therefore potentially at a lower interest rate.

Transaction costs

The transaction costs involved in purchasing the land must be factored into the residual value of the site. In addition, when assessing the GDV, transaction costs, if appropriate, must also be borne in mind.

Example: Hotel Butler

In this example, the Hotel Butler is anticipated to trade as shown in Table 12.1.

Equal consideration is then required of the development costs as shown in Table 12.3.

It is important that these building costs are analysed to ensure that they are appropriate for the sort of development that is proposed, and indeed that the building costs are typical of the sort of costs that would be expected within the market.

Any savings that a particular developer can bring to a development that are not typical of the market need to be discounted when carrying out the residual valuation (if the purpose of the exercise is to discover the market value).

Analysis of the building costs is normally carried out on a per key, or per square metre, basis, as shown in Table 12.4.

In this example, contingency allowances have been specifically allocated in various parts of the equation which total just under 8.2 per cent of the final development costs.

It is agreed that the finance costs have been taken at 6.5 per cent per annum in this instance, which is the prevalent borrowing rate for developments of this type. In this case, it has been assumed that due to the phase of the development programme, the borrowing requirement will equate to half of the money being needed at the start of the development process, with the other half being required at a later stage.

Developer's profit

This is a contentious item, as potentially each developer has their own different development criteria for carrying out developments.

Certain hoteliers are less concerned with development profits than with getting new hotels built, which effectively reduces the amount of development profit that the valuer needs to incorporate into the valuation.

However, it will be down to the experience of the valuer to determine what level of developer's profit to deduct.

Land value

To determine the value of the site, it is important that the valuer deducts the cost of financing the purchase from the day when such costs will accrue, and includes the transaction costs to arrive at the net site value.

It should be reiterated that this method is very susceptible to minor changes in the variables and as such, the valuer would do well to place as much emphasis on comparable transactions as possible.

In the example of the Hotel Butler, the site value equated to £2,500,000, which breaks back to £11,364 per room.

Table 12.1 Hotel Butler trading projections

	Forecast in present values					
Year	*1*		*2*		*3*	
No. of rooms	220		220		220	
Rooms available	80,300		80,300		80,300	
Rooms sold	48,180		56,210		60,225	
Occupancy	60.0%		70.0%		75.0%	
ADR (£)	80.00		90.00		100.00	
RevPar (£)	48.00		63.00		75.00	
Growth (RevPAR)			31.3%		19.0%	
Revenues (£ '000s)						
Rooms	3,854.4	70.0%	5,058.9	67.5%	6,022.5	65.0%
Food and beverage	1,211.4	22.0%	1,836.2	24.5%	2,501.7	27.0%
Leisure club	330.4	6.0%	449.7	6.0%	555.9	6.0%
Other revenue	110.1	2.0%	149.9	2.0%	185.3	2.0%
Total revenue	5,506.3	100.0%	7,494.7	100.0%	9,265.4	100.0%
Departmental profit (£ '000s)						
Rooms	2,505.4	65.0%	3,642.4	72.0%	4,456.7	74.0%
Food and beverage	363.4	30.0%	642.7	35.0%	1,000.7	40.0%
Leisure club	79.3	24.0%	107.9	24.0%	133.4	24.0%
Other revenue	35.2	32.0%	48.0	32.0%	59.3	32.0%
Total departmental profit	2,983.3	54.2%	4,441.0	59.3%	5,650.0	61.0%
Departmental costs (£ '000s)	2,523.0	45.8%	3,053.7	40.7%	3,615.4	39.0%
Undistributed operating expenses (£ '000s)						
Administrative and general	550.6	10.0%	674.5	9.0%	787.6	8.5%
Sales and marketing	330.4	6.0%	374.7	5.0%	370.6	4.0%
Repairs and maintenance	220.3	4.0%	299.8	4.0%	370.6	4.0%
Utility costs	176.2	3.2%	239.8	3.2%	296.5	3.2%
Total undistributed expenses	1,277.5	23.2%	1,588.9	21.2%	1,825.3	19.7%
Income before fixed charges (£ '000s)	1,705.8	31.0%	2,852.1	38.1%	3,824.8	41.3%
Fixed charges (£ '000s)						
Reserve for renewals	110.1	2.0%	224.8	3.0%	370.6	4.0%
Property taxes	258.8	4.7%	352.2	4.7%	435.5	4.7%
Insurance	55.1	1.0%	74.9	1.0%	92.7	1.0%
Management fees	165.2	3.0%	224.8	3.0%	278.0	3.0%
Total fixed charges	589.2	10.7%	876.9	11.7%	1,176.7	12.7%
EBITDA (£ '000s)	**1,116.7**	20.3%	**1,975.2**	26.4%	**2,648.0**	28.6%

Table 12.2 Hotel Butler income capitalisation

Net base cashflow			
EBITDA in present values in year 3		£2,648,047	
Capitalised at 9.25%		10.81	£28,627,534
less			
	Income shortfall		£2,204,200
	Subtotal		£2,204,200
Gross value			£26,423,335
Say			£26,400,000

It is at this point that the valuer needs to stand back and see if the end result is sensible.

Reference needs to be made to any other site transactions that are relevant. It is sometimes useful to review the land value as a proportion of the overall end costs and value. In this instance, £11,364 per room equates to approximately 8.1 per cent of the development cost and 6.4 per cent of the end value, which is deemed normal in this particular area. The proportion of the overall value that the land value equates to will normally relate directly to the scarcity of supply of suitable sites, with hotel developments in Hong Kong showing a higher element of land value as a percentage of the completed development than for a coastal resort hotel in Kenya.

Table 12.3 Hotel Butler development costs

	£ Total		£
Value at completion			26,400,00
Construction costs			
Structure, shell and core	4,925,123		
Internal secondary works	1,788,555		
M&E works	4,021,000		
Fitting-out works	345,000		
Finishes	680,000		
External works	100,000		
Preliminaries	2,555,125		
Connection to utilities	146,250		
Contingency	3,124,025		
Subtotal	17,685,078		
Construction fees and insurance			
Architect	350,000		
Employer's agent	70,000		
Project management	115,400		
Insurance and la fees	105,250		
Contingency	42,150		
Subtotal	682,800		
FF&E	The associated costs to be met by the operator		
Pre-opening costs	200,000		
Subtotal	18,567,878		
Other expenses and fees			
Operator's technical assistance fees	152,500		
Legal fees	62,500		
Financial costs	1,250,000		
Subtotal	1,465,000		
Land costs			
Rights of access insurance	100,000		
Legal, surveying and planning consultants' costs	300,000		
Subtotal	400,000		
Total development cost	20,432,878	£ 92,877	per bedroom
Finance costs	6.5%	664,069	
Developer's profit	15.0%	3,064,932	
Total costs		24,161,878	
Gross site value			2,238,122
Net of stamp duty	4.0%		2,152,040
Site finance @ 6.5% for 1.5 years			218,217
Stamp duty			86,082
Site value			1,933,823
Say			1,935,000
Per bedroom			9,258

Table 12.4 Hotel Butler building costs

Bedrooms: 220

Size: 12,120m²

Construction costs	*£ Total*	*£/Key*	*£/m²*	*% of total construction costs*	*% of overall cost*
Structure, shell and core	4,925,123	22,387	406	27.85%	20.38%
Internal secondary works	1,788,555	8,130	148	10.11%	7.40%
M&E works	4,021,000	18,277	332	22.74%	16.64%
Fitting-out works	345,000	1,568	28	1.95%	1.43%
Finishes	680,000	3,091	56	3.85%	2.81%
External works	100,000	455	8	0.57%	0.41%
Preliminaries	2,555,125	11,614	211	14.45%	10.58%
Connection to utilities	146,250	665	12	0.83%	0.61%
Contingency	3,124,025	14,200	258	17.66%	12.93%
Subtotal	17,685,078	80,387	1,459	100.00%	73.19%
Construction fees and insurance				*% of total construction fees and insurance*	*% of overall cost*
Architect	350,000	1,591	29	62.30%	2.28%
Employer's agent	70,000	318	6	7.93%	0.29%
Project management	115,400	525	10	13.07%	0.48%
Insurance and la fees	105,250	478	9	11.92%	0.44%
Contingency	42,150	192	3	4.77%	0.17%
Subtotal	682,800	3,104	57	100.00%	3.65%

FF&E	The associated costs to be met by the tenant				
Pre-opening costs	200,000				
Subtotal	18,567,878				76.85%
Other expenses and fees					
Operator's technical assistance fees	152,500				
Legal fees	62,500				
Financial costs	1,250,000				
Subtotal	1,465,000				
Land costs				*% of total land costs*	*% of overall cost*
Rights of access insurance	100,000	455	8	25.00%	0.41%
Legal, surveying and planning consultants' costs	300,000	1,364	25	75.00%	1.24%
Subtotal	400,000	1,818	33	100.00%	1.66%
Total development cost	20,432,878	92,877	1,686		84.57%
Finance costs	6.5%	664,069		17.81%	2.75%
Developers profit	15.0%	3,064,932		82.19%	12.68%
Total costs		**24,161,878**			**100.00%**

13 Serviced apartment values and how they are determined

Introduction

This chapter will outline the different types of serviced apartments, how they work as businesses, why they appeal to customers, why they appeal to investors and how they are valued.

Over the last forty years in North America and over the last twenty years in Europe and Asia, a specialist category of guest accommodation has been developing and growing in popularity, both with customers and with investors. The serviced apartment category typically provides an apartment rather than a hotel room, sometimes with limited other services, and it tends to be aimed at the longer-stay market. It should be noted that suite-hotels do not tend to exhibit the same trading profit, being more closely linked to traditional hotels. That said, some upper-upscale serviced apartments operate almost exactly like hotels in the level of services they provide their customers, with everything from room service, private dining and concierge services.

The serviced apartment market is quite a specialist subsegment of the overall hospitality industry, but at heart it is based on the same principles as the wider hotel investment market. If the right product is in the right place and well run, providing good service and value for money, then in most instances it will be a successful business.

All the factors that apply to the hotel market also apply to the serviced apartment market when it comes to running a successful business. The quality of the staff, the ability to anticipate market needs, effective marketing, generating word of mouth and repeat business, along with careful segmentation of demand is just as important for serviced apartments as it is for traditional hotels.

The key differences are that a customer tends to feel they are getting better value for money in a serviced apartment (certainly in terms of the size compared with the cost to stay), whilst the reduction of services offered tends to generate higher returns on investment for investors.

However, the serviced apartment market is not one simple category of 'hotel', and indeed there is often confusion as to what constitutes a serviced apartment. As such, it is important to understand the various different types of serviced apartments that are available in the market, so the various differences can be understood, and the owner and valuer can understand the factors that relate to each type.

Reinvest to remain the best!

Nick Newell, director – Savills (Hotels & Leisure)

Hotel value is ultimately a factor of yield and operational performance (EBITDA), and it is therefore important that the operator can maintain profitability. In the face of increasing supply, it is imperative to maintain quality through regular reinvestment into the hotel product.

In many instances, we witness the demise of a once-grand hotel with trading performance suffering as a consequence of a lack of investment. Quite often, this can lead to a failure to meet loan covenants and in extreme situations can lead to the hotel being placed into administration and sold at a discounted price.

In assessing the operational performance achievable by a hypothetical efficient operator, it is important to ensure that the reinvestment requirements are adequately allowed for to meet these projections. The typical method is to allow for a hypothetical amount to be spent on average each year to maintain the hotel, referred to as the 'fixtures fittings and equipment reserve' (FF&E reserve). Typically, it is prudent to allow for an amount equating to 3 per cent of total revenue p.a. in a stabilised year of trading for a budget hotel, 4 per cent of total revenue p.a. for a mid-market hotel and 5 per cent of total revenue p.a. for a luxury/five-star hotel.

In reality, hotel management will undertake a phased programme of works to replace fixtures and fittings in order to cause minimal disruption to the operation of the hotel. A hotel might typically be refurbished on a five to seven-year cycle. However, for the valuer, it is important not to make 'knee jerk' deductions in capital expenditure for refurbishment in any given year, and ultimately it is fair to assume that an annual amount would be deducted and set aside in a reserve fund to spend when appropriate to maintain the hotel and protect its profitability and operational performance.

Different types of serviced apartment complexes

Serviced apartments are a hybrid of two property classes, hotels and residential property. The 'combined entity' provides an alternative to one or both property classes for investors. However, the term 'serviced apartment' is not particularly well understood as it covers a number of subcategories of property, as described in the following pages.

Serviced apartments are positioned to meet the need of customers that are looking for flexible accommodation options. Their customers typically have longer length of stay requirements and are after more space than is typically offered by traditional hotels.

Figure 13.1 Serviced apartment in Morocco

Branded residences

Branded residences started as the residential components of multi-ownership resorts but have changed over the years into a wider category of products. Now many branded residences are an extension of adjacent hotels, owned by the same party.

In essence, the residential property (whether an apartment or a villa) is operated by an adjacent hotel operator, and is branded by the hotel operator. Some of these units (when the overall property is in multi-ownership) may be kept for the exclusive use of the owner, in which case the brand merely offers services to the property owners and ensures the maintenance and cleanliness of the property.

However, most are operated as letting units, in the same way a hotel is let, with guests staying for an agreed length of time, anywhere from between one night and a year. There has been a recent trend towards branded residences being concentrated towards the upper-upscale segment of the hotel market, typically operated by high quality hoteliers like Four Seasons, Ritz Carlton, Mandarin Oriental or Jumeriah, who, as such, fit into the illusive 'lifestyle' concept where experience is more important than price, attracting a relatively wealthy customer base.

A typical branded residence can be anywhere from a studio apartment, to a five-bedroom apartment in a city centre complex, through to a seven-bedroom villa in a resort complex. All facilities are usually provided by the adjacent hotel upon request, and at an extra charge.

Figure 13.2 Kitchen in a serviced apartment

Branded residences have grown in popularity for developers for a number of reasons:

- profit margins are high compared with standard hotel returns;
- sales prices tend to have a premium of between 10 per cent and 30 per cent on comparable residential prices;
- the possibility of selling off individual units allows for multi-ownership funding options.

Typically, branded residences will be located in prime city centres or in high-end resorts, where the demand for the residential product will be high enough to justify the development.

Corporate housing

Corporate housing is a product that originally evolved from the short-term apartment rental market. Typically, it is used as a short-term measure for corporate employees who are being relocated, allowing them to occupy a property for a period until they can arrange their own purchase or letting of a new home. Many countries have established regulations to govern this market, enforcing minimum stay periods (thiry to niety days) to minimise disruption to surrounding tenants.

Corporate housing complexes can sometimes actually be a wide selection of properties in different locations, rather than in one homogenous block, with

Hotel branded residences

Hotel branded residences (HBR) are a growth area in the residential property market, often providing significant premiums on comparable residential properties that are not aligned with a hotel brand.

The concept is not new, having evolved from the days when hotels attracted long-term residents, like Coco Chanel who lived in The Ritz in Paris for over thirty years, Peter Sellers who lived in The Dorchester in London for decades and Cole Porter who spent his last fifteen years living in the Waldorf Astoria in New York. The Sherry-Netherland Hotel is sometimes credited with being the first example of hotel branded residences, opening in New York with a mix of hotel rooms and permanent residential apartments.

In recent years the concept has been refined and tends to be most prevalent in the luxury end of the market, with operators such as Four Seasons, Hyatt, Ritz Carlton, Aman Resorts, W Hotels, Rosewood, Fairmont, Mandarin Oriental, Jumeriah and St Regis some of the market leaders worldwide.

HBR are typically private residences that have been developed for sale to individual owners, and can be used exclusively by that owner. This is in contrast to typical hotels or aparthotels where they are usually built for short-term lets. Sometimes, where the HBR are developed in conjunction with a hotel, these residences can be let by the hotel operator, providing a potential income for the individual owner who has still purchased the property, and owns it exclusively.

The attractions of HBR for 'owners' is primarily the opportunity to own a premium residence associated, and potentially serviced, by the hotel. These services can include all typical luxury hotel features, including:

- twenty-four-hour concierge services
- limousine/chauffeur/transport services
- spa facilities
- F&B facilities, including room service
- housekeeping
- property maintenance
- security
- business centre services
- gym facilities, including personal training
- babysitting/childcare services
- function and event planning.

A limited number (as an overall percentage) purchase such properties as investments, where the model allows for the residence to be let to 'hotel guests'.

The attraction for the developer is the premium above-standard residential prices that such an association with the hotel adds to the sales price.

The attraction for the operator is usually the fees associated with such developments, although with the right project it can help enhance brand association for the hotel brand itself. Fees are typically earned from management fees where the HBR (or a proportion of them) form part of the letting inventory.

Service charge fees will also be earned where the HBR are serviced by the hotel. In addition, some agreements allow for a percentage of the sale price of the HBR to be paid to the operator, for the association with the brand during the marketing of the property.

A typical ownership map of a combined hotel and HBR product will have the hotel operator contracting with only one partly to provide the management services, and as such some form of single entity (like a residence owners' entity).

Discussing the valuation of such units in the most part is beyond the scope of this book, as they are effectively just residential valuations, based upon comparable evidence in that location, taking into account the premium the market is prepared to pay for the 'brand association'.

Premiums above residential sales prices range significantly depending on the scheme, the brand and the market-demand, from 5 per cent–50 per cent in certain situations.

In Chapter 14 the impact HBR can have upon resort values and site values where they form part of the fractional ownership scheme has been addressed.

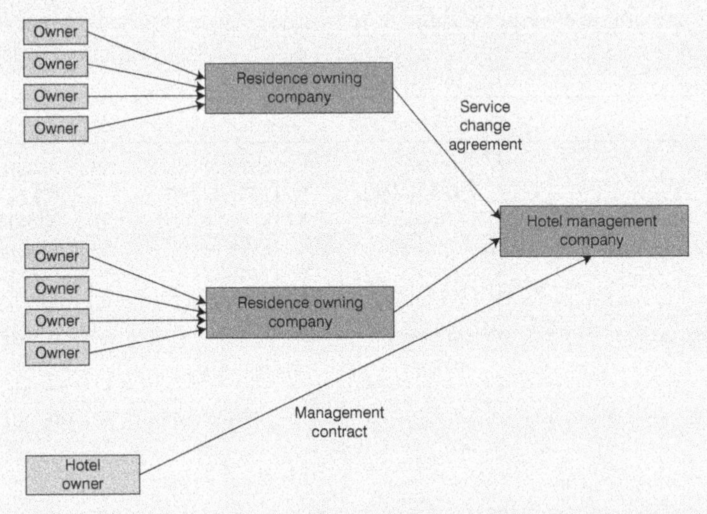

Figure 13.3 Typical ownership structure for hotel branded residences as part of a mixed-use scheme

100 units being split up into as many as twenty different buildings. This can impact on the operational efficiency of the overall business.

A typical corporate housing unit will comprise a full fitted apartment, ranging from studio to a four-bedroom apartment, with separate bathrooms, kitchen and lounge. In addition, many will also have designated car parking spaces.

Aparthotels

Of all types of serviced apartments, aparthotels are closest to a traditional hotel, albeit there are still many significant differences. Sometimes referred to as extended-stay hotels, aparthotels are self-catering establishments that are serviced. The rooms are typically cleaned on a weekly basis, and there is a twenty-four-hour reception desk and usually an on-site laundry for guest use. Some aparthotels have bars, restaurants and even meeting rooms, similar to hotels. Usually breakfast is provided.

The aparthotel is typically contained within a single building, similar to a traditional hotel or apartment block.

Most aparthotel accommodation ranges from studio apartments to three-bedroom apartments, typically providing a bedroom unit with en suite bathroom and a combined lounge with kitchenette area.

Aparthotels tend to be discounted from comparable hotel rooms, despite typically being 20–30 per cent larger, and pricing tends (on a per night basis) to drop when longer stays are booked.

Basic valuation methodology

As with all trading properties, the nature of each business can be so varied that direct comparable evidence is difficult to find. A serviced apartment building will

Table 13.1 Differences between serviced apartment categories

	Branded residences	*Corporate housing*	*Aparthotel*
Facilities	• Any size • Full kitchen • Lounge/diner • Work area • Laundry • Car parking	• Any size • Full kitchen • Lounge/diner • Work area • Laundry • Car parking	• Studio, 1 or 2 bed • Kitchenette • Lounge/diner • Onsite laundry
Services	• All hotel services • Reception at hotel • Daily cleaning	• Weekly cleaning	• Some minor services • Reception at hotel • Daily cleaning
Average length of stay	1 day–1 year	30 days–2 years	2–5 days
Market segment	Upper-upscale	Upper-upscale to mid-market	Budget to upscale

have its own business model and customer base, which is usually quite unique to that property. In addition, the lack of serviced apartments, and the lack of transactions on the open market generally, tends to lead to a lack of comparable evidence that valuers can rely upon.

As such, the preferred method of valuation tends to be the 'profits method', although it must be stressed that comparable evidence is still important for each part of the valuation under the profits method. Comparable transactions where available will help define yield selection, whilst comparable trading profiles will help determine potential trading profitability for the property.

Example: Swann Suites

In this example this aparthotel has 150 letting units (a mix of studios and 1-bedroom apartments), all available for let on a daily basis.

It is trading in the mid-market section and revenue generation is dominated by the rooms department. Breakfast is the main source of F&B income, charged separately from room hire, whilst other revenue comprises vending machine sales, car parking charges and income from premium Wi-Fi connections.

Occupancy is anticipated to grow from 82.5 per cent to a stable level of 85 per cent in the second year of trading, whilst ADR will grow over three years to a stable level of CHF155.92, generating a stable RevPAR of CHF132.53.

Rooms generate 85–86 per cent of total revenue, and with the high profit margins generated by rooms, the IBFC converts at 57.4 per cent in the stable years of trading.

The property is valued using a 7.25 per cent capitalisation rate, reporting a gross value of CHF51m, equating to CHF340,000 per bedroom.

Operational challenges – getting to the bottom line

When carrying out the valuation of the serviced apartments, getting to the sustainable EBITDA of the property is the immediate goal. Similar to a standard hotel, this is usually done through five quite separate exercises:

1. property inspection
2. analysis of the historic trading accounts
3. review of the competition
4. interview with the manager
5. independent analysis of everything that has been seen and said throughout the process to determine how the property is likely to trade in the hands of a reasonably efficient operator (REO).

Assessing the likely trading profile of the serviced apartment will require a complete understanding of the trading environment in which the facility sits.

It is difficult to provide a comprehensive check list of things that need to be considered, as each serviced apartment complex tends to be quite different.

Table 13.2 Swann Suites trading projections

	CHF		CHF		CHF	
	Year 1		*Year 2*		*Year 3*	
No of rooms	150		150		150	
Rooms available	54,750		54,750		54,900	
Rooms sold	37,482		37,914		38,270	
Occupancy	82.5%		85.0%		85.0%	
ADR (CHF)	145.00		152.12		155.92	
RevPAR (CHF)	119.63		129.30		132.53	
Growth (RevPAR)			8.1%		2.5%	
Revenues (CHF '000s)						
Rooms	6,549.5	85.0%	7,079.3	86.0%	7,256.3	86.0%
Food and beverage	770.5	10.0%	823.2	10.0%	843.8	10.0%
Other revenue	385.3	5.0%	329.3	4.0%	337.5	4.0%
Total revenue	7,705.3	100.0%	8,231.7	100.0%	8,437.5	100.0%
Departmental profit (CHF '000s)						
Rooms	5,567.0	85.0%	6,017.4	85.0%	6,167.8	85.0%
Food and beverage	269.7	35.0%	288.1	35.0%	295.3	35.0%
Other revenue	173.4	45.0%	148.2	45.0%	151.9	45.0%
Total departmental profit	6,010.1	78.0%	6,453.7	78.4%	6,615.0	78.4%
Undistributed operating expenses (CHF '000s)						
Administrative and general	500.8	6.5%	535.1	6.5%	548.4	6.5%
Sales and marketing	269.7	3.5%	288.1	3.5%	295.3	3.5%
Repairs and maintenance	385.3	5.0%	411.6	5.0%	421.9	5.0%
Utility costs	462.3	6.0%	493.9	6.0%	506.3	6.0%
Total undistributed expenses	1,618.1	21.0%	1,728.7	21.0%	1,771.9	21.0%
Income before fixed charges (CHF '000s)	4,392.0	57.0%	4,725.0	57.4%	4,843.1	57.4%
Fixed charges (CHF '000s)						
Reserve for renewals	0.0	0.0%	0.0	0.0%	0.0	0.0%
Property taxes	346.7	4.5%	370.4	4.5%	379.7	4.5%
Insurance	69.3	0.9%	82.3	1.0%	84.4	1.0%
Management fees	0.0	0.0%	0.0	0.0%	0.0	0.0%
Total fixed charges	416.1	5.4%	452.7	5.5%	464.1	5.5%
EBITDA	**3,975.9**	**51.6%**	**4,272.3**	**51.9%**	**4,379.1**	**51.9%**

Table 13.3 Swann Suites income capitalisation (in CHF)

Net base cashflow			
EBITDA in present values in year 3		4,379,073	
Capitalised at 8.50%		11.765	51,518,500
Less	Income shortfall	509,966	
	Capital expenditure		509,966
Gross value			51,008,534
SAY			51,000,000
Price/apartment			340,000

However, by breaking down the trading profile into sections is usually desirable, looking at the various income streams, followed by the various operational costs, is one of the most sensible approaches as it will usually result in a thorough analysis of the likely vagaries of that specific business.

Income

Typically, a serviced apartment will generate the vast majority of its revenue through the accommodation lettings. Income from this source will be influenced by room categories and configuration, and how well the letting inventory meets the local customers' requirements.

There are a number of serviced apartments that offer a wide range of additional services that can also generate additional income. These may include:

- car parking
- meeting room hire
- F&B income
- telephone/Wi-Fi charges
- concierge services
- office services (printing, secretarial services, etc.)
- food parcels
- rental income
- retail.

Costs

Costs can be broken down into many areas including the direct cost of sales, staffing costs, sales and marketing, laundry, administration and general, repair and maintenance, water charges, insurance and taxes.

Australian serviced apartments – strata ownership

Adrian Archer – Savills Australia

Australia has a successful and mature serviced apartment market in comparison to many countries. The success of the expansion of this sector has been driven by the ability to sell individual apartments in a complex, known as strata title. Apartments are primarily used as holiday homes, generally providing a form of income whilst not owner occupied. A strata manager is engaged to lease out the apartments on behalf of the owners on a nightly basis similar to a hotel. In return, strata managers receive a fee for this service, known as management rights.

Returns are attractive compared to more traditional residential investments, and apartment values usually exceed the going concern value of the property 'in one line'. Large national operators, such as Mantra, Oaks and Quest, require their apartments to meet certain brand standards. Obtaining funds from all the individual apartment owners to coordinate a consistent renovation inside each apartment can be a long and difficult process. If brand standards are compromised, the quality of the apartments can become inconsistent. Ultimately, this damages the brand, guest experience, room rates, investors' return and eventually value.

The strata model is popular in resort locations, such as the Gold Coast. This popularity is driven by the attractiveness of a lifestyle investment choice by investors. The strata model also provides a value uplift in comparison to the 'in one line' model.

'In one line' model is characterised by all apartments and public areas in a complex remaining under the ownership of a single investor. The apartments and business are valued on a cashflow basis as opposed to on an individual apartment basis. Central business district locations in major cities favour the 'in one line' operating model, providing the owner with more flexibility to refurbish the apartments as and when required. City locations are usually more immune to the seasonal fluctuations of resort locations and therefore provide more consistent cashflow. Upon sale of the asset, a purchaser may have the ability to choose to obtain vacant possession of the entire property to rebrand or redevelop, subject to the terms of the operators' management agreement.

Serviced apartment developers need to consider the location of each development on its merits before deciding which model to adopt. Short-term gains in strata title can be at the detriment of long-term investment returns. The long standing rule that hotel development is a long-term investment still stands true today.

Inspection

A typical serviced apartment inspection will follow the general course of a hotel inspection. Initially, the valuer will request the information required, ahead of the inspection.

This will include all, or some, of the following.

Legal information

- title documentation, or reports on title
- ground leases or property leases
- subleases
- occupational licences or concessions
- planning consents for the site and building(s) on the property
- any restrictive covenants and other use prohibitions
- details of any unused planning consents
- details of any unfulfilled planning conditions
- copies of the operating licences.

Property information

- area of the overall site;
- site plans (with sizes) for each building;
- building plans – floor by floor;
- breakdown of apartment facilities (letting inventory by type, category, size and quantum);
- breakdown of other facilities including F&B outlets;
- copies of any current or historic valuation reports;
- copies of any current or historic condition surveys, asbestos surveys, etc.;
- details of historic capital expenditure on the properties (last three years);
- details of any proposed capital expenditure (this year and next year's budget).

Operational information

Historic profit and loss accounts (3 years minimum). These will be the full P&L accounts, not just the summary sheet:

- management statistics, including occupancy, ADR, RevPAR and average length of stay;
- statutory accounts;
- full P&L budget for the current year and the next year;
- details of market segmentation and geographical analysis of the guest profile;
- staff details – numbers of staff and contracts of employment, including pension details, length of service and skills training for each member of staff;
- sales and marketing plans; this year and next year's budget

- details of any F&B revenues; where it comes from, average cover spends, etc.;
- details of the local markets; SWOT analysis of the local competitors;
- details of local trading; how does the property compete with its competitive set?
- details of operational equipment.

Other

- details of any outstanding complaints, legal claims, etc.
- details of the tax positioning of the property (outstanding tax bills, capital allowances, etc.).

The serviced apartment complex will be inspected, usually in the presence of the general manager. The valuer will be looking at the property to see that it is suitable for use and how appealing it will be to the local market.

This will relate to the layout of the facilities, the quality of the accommodation, as well as the design and suitability of any other facilities.

In addition, the valuer will be noting details about the following:

- Car parking – is it suitable for the volume of customers? Could it handle additional business without creating operational difficulties? Is there spare capacity that could be used for an alternative use? Could any spaces be let out to third-party occupiers from nearby commercial/residential properties?
- Non-income-generating space – are there areas that could be reconfigured into revenue-generating spaces? Are the existing public areas suitable for the current and future operation of the business?
- Rights of way – are there any public rights of way? Do they impinge on the operation of the business?
- Other facilities – are there any other areas that could generate income for the business? Are there any items that might raise costs for the business?
- Equipment – is all the equipment up to date and suitable for use? Does it have operational issues that need to be reflected in the trading profile? Will it need to be replaced soon?
- Health and safety – does the property comply with current and future legislation? If not, what are the cost implications?

In effect, the purpose of the inspection is to see all matters that may impact on revenue generation or have cost implications, so an accurate assessment of future trading can be made.

The condition of everything will also be assessed to ensure that any future capital expenditure requirements can be accurately gauged.

The quality of the property is then assessed, in relation to the relevant market, so an accurate yield for the property can be ascertained to finally determine the value.

Interview with the general manager

After the inspection of the property, the trading history, past, present and future, will be discussed with the management.

Typically, a detailed discussion will review the operational issues relating to the serviced apartment complex in terms of revenue generation, followed by anticipated costs resulting in an understanding of the future trading profile and a review of the proposed capital expenditure plans.

The local market can be key to revenue generation for the property, with new supply potentially impacting on income levels. As such, a detailed discussion with the general manager about the various strengths and weaknesses of the property, in relation to the existing (and proposed) market, is desirable to allow the valuer to understand the local market dynamics as seen from the perspective of the current manager.

The income generated by each apartment/room category, including segmentation of demand, occupancy levels and ADRs, will be reviewed in light of past trading so an understanding of the projections can be fully made.

The same process will occur for each revenue line, followed by each cost line to ensure the trading projections are typical of the local market, and achievable.

Example: Aparthotel Pollock

In this example, the luxury serviced apartment complex has eighty-five letting units. It offers a great range of additional services to guests, including stocking cupboards with food, personal chefs, concierge services, gym and spa. It is in a well-developed and established market (Table 13.4).

Occupancy rates are high, starting at 82.5 per cent and anticipated to stabilise at 85 per cent, based partly on the prevalence of long lets in the local market. The ADR is high, based on the positioning within the market and the apartment mix within the letting inventory.

Only 65–66 per cent of all income is generated by rooms, with F&B revenue generating 12.5 per cent and other services contributing 21.5 per cent to 22.5 per cent of turnover.

The nature of the revenue mix has resulted in a relatively low EBITDA, ranging between 31.4 per cent and 31.8 per cent.

The property is valued using a 7.25 per cent capitalisation rate, reporting a gross value of €71.3m, equating to €838,824 per apartment. This breaks back to around €2,750 per sq. ft. which is appropriate for similar residential property values.

Table 13.4 Aparthotel Pollock trading projections

	Year 1		*Year 2*		*Year 3*	
No. of rooms	85		85		85	
Rooms available	31,025		31,025		31,110	
Rooms sold	25,596		26,371		26,444	
Occupancy	82.5%		85.0%		85.0%	
ADR (€)	355.00		400.00		412.50	
RevPAR (€)	292.88		340.00		350.63	
Growth (RevPAR)			16.1%		3.1%	
Revenues (€ '000s)						
Rooms	9,086.4	65.0%	10,548.5	66.0%	10,878.1	66.0%
Food and beverage	1,747.4	12.5%	1,997.8	12.5%	2,060.3	12.5%
Other revenue	3,145.3	22.5%	3,436.3	21.5%	3,543.6	21.5%
Total revenue	13,979.1	100.0%	15,982.6	100.0%	16,482.0	100.0%
Departmental profit (€ '000s)						
Rooms	7,269.2	80.0%	8,438.8	80.0%	8,702.5	80.0%
Food and beverage	498.0	28.5%	569.4	28.5%	587.2	28.5%
Other revenue	943.6	30.0%	1,030.9	30.0%	1,063.1	30.0%
Total departmental profit	8,710.8	62.3%	10,039.1	62.8%	10,352.8	62.8%
Undistributed operating expenses (€ '000s)						
Administrative and general	1,188.2	8.5%	1,358.5	8.5%	1,401.0	8.5%
Sales and marketing	629.1	4.5%	719.2	4.5%	741.7	4.5%
Repairs and maintenance	699.0	5.0%	799.1	5.0%	824.1	5.0%
Utility costs	838.7	6.0%	959.0	6.0%	988.9	6.0%
Total undistributed expenses	3,355.0	24.0%	3,835.8	24.0%	3,955.7	24.0%
Income before fixed charges (€ '000s)	5,355.8	38.3%	6,203.2	38.8%	6,397.1	38.8%
Fixed charges (€ '000s)						
Reserve for renewals	0.0	0.0%	0.0	0.0%	0.0	0.0%
Property taxes	838.7	6.0%	959.0	6.0%	988.9	6.0%
Insurance	125.8	0.9%	159.8	1.0%	164.8	1.0%
Management fees	0.0	0.0%	0.0	0.0%	0.0	0.0%
Total fixed charges	964.6	6.9%	1,118.8	7.0%	1,153.7	7.0%
EBITDA (€ '000s)	**4,391.2**	**31.4%**	**5,084.5**	**31.8%**	**5,243.3**	**31.8%**

Table 13.5 Aparthotel Pollock income capitalisation (in €s)

Net base cashflow			
EBITDA in present values in year 3		5,243,346	
Capitalised at 7.25%		13.793	72,322,016
Less	Income shortfall	1,011,035	
	Capital expenditure		1,011,035
Gross value			71,310,981
SAY			71,300,000
Price/apartment			838,823

14 Resorts with fractional ownership units and how their value is determined

Introduction

Over the last fifty years, a large number of resorts have been developed that combine leisure facilities with residential properties, to meet the needs of the holidaymakers using the facilities. Apartments and villas, as opposed to traditional hotel rooms, tended to suit larger families, and made their stay more comfortable. As such, these resorts have become quite popular with guests, typically leading to better overall financial performance.

To help fund new resort developments, a number of developers started to sell off the residential components of the resort to third-party owners. This provided a cashflow for the development, allowing it to progress.

In recent times, there has been a sharp rise in mixed-use resorts where part of the guest accommodation on offer to holidaymakers is actually private residences owned by third parties. Sometimes these units are operated as 'hotel rooms' by the resort and other times they are kept separate from the letting inventory. However, as long as the upkeep and visual impact of the private residences is kept at a suitable level, this can be a highly complementary use from the hotel owner's (and indeed hotel operator's) perspective. The 'non-hotel guests' can still use the bars and restaurants, which generates additional revenue for the resort.

It is beneficial for the third-party residential owners as well, as they get to benefit from having all the facilities of a resort on their doorstep. Then developers of such schemes started to widen the 'buying pool' by working on a concept where the hotel manager would sell and service the private apartments when they were not being used, allowing investors (and quasi-investors) to become interested in the product.

The methodology of this style of funding has evolved over the years, with ownership problems and operational issues potentially arising from most structures if suitable safeguards are not put in place at the start of the project.

This chapter outlines some of the key structures and addresses some of the issues that need to be safeguarded against when structuring such multi-ownership projects. It also outlines how the valuer should approach the valuation of such an entity.

This chapter will not detail how to value the individual villa or apartment (or fractional or timeshare interest in a single unit), as that will be led almost exclusively by comparable transactions (whether new sales or resales), with amenity value and the various structures, and future cost implications, making the value relatively simple to assess.

Instead, it will address assessing the value of the whole of the resort, taking into account the fractional 'income streams and costs' that are allocated to the overall resort, outlining how various factors will impact on the overall value of the resort. However, it is important that the valuer understands the benefits and shortfalls of the potential structures that are commonly in place in the market, so they can, when necessary, assess the value of any unsold units as part of the overall property.

What is a multi-ownership resort and why are they created?

Any resort where there is more than one owner is technically a multi-ownership resort, though for the purposes of this section we are talking about schemes that involve multiple residential units that are owned or part owned by a number of different people.

All of the above examples have been used to develop resorts, with possibly the most important being the fractional ownership market and the single owner-per-unit (SOPU) ownership market. All the other types of ownership structure are just variants upon these two key types.

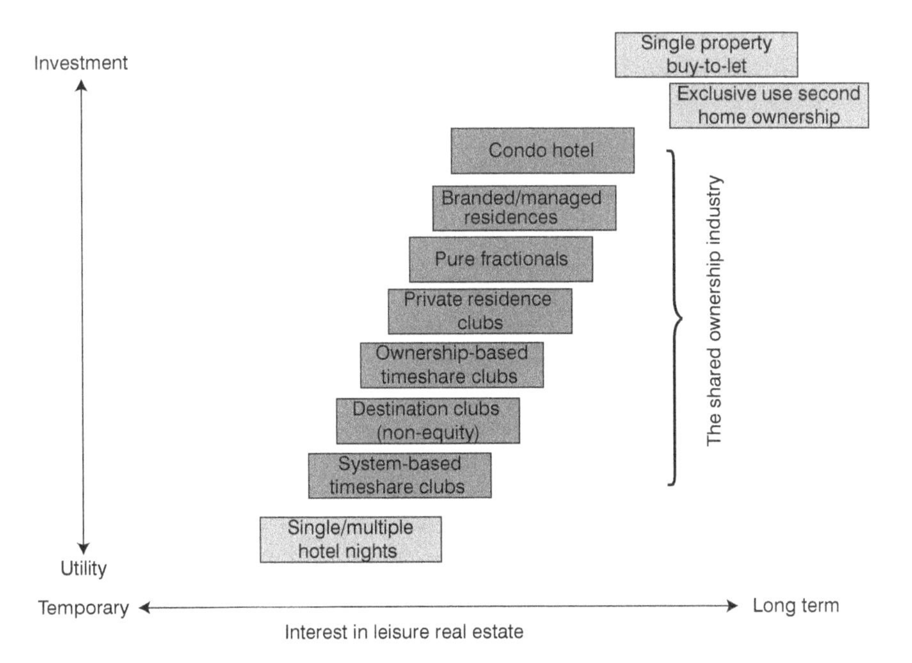

Figure 14.1 The shared ownership industry

Fractional ownership

One of the earliest examples of multi-ownership resorts were known as timeshares, where one entity purchased the right to use a certain property (or part of a property) for a certain time each year.

Timeshares developed a very bad reputation, as high-pressure sales techniques were commonly used to persuade people to invest in properties with the lure of 'pre-paid' vacations for the rest of their lives. However, the truth of such schemes was sometimes very different from what was advertised. The key issues tended to be that some timeshares were difficult to use, had undesirable facilities and could even be more expensive to use than regular hotels. Selling such properties became difficult, and even just giving up the property was difficult, as maintenance contributions were liabilities signed up to for life.

In response to the adverse publicity surrounding timeshares, traditional investment across large parts of the world tended to be concentrated into second homes for a number of years, with one party owning the whole of the property and letting it out (either on their own or with the help of the resort operator) when it was not being used.

After years of bad publicity, timeshares were mostly replaced by fractional ownership schemes, which have a much better perception in the investment market, even though they tend to be very similar legally.

The most important single difference between timeshares and modern fractional ownership properties is the number of owners for each property. Traditional timeshare properties had typically somewhere between twenty-siz to fifty-two 'owners', as they were sold on one- or two-week blocks. As such there was little 'pride of ownership' and this lack of connection translated apathy into a lack of care. Higher traffic also meant more wear and tear.

By contrast, modern fractional properties have a much lower number of owners, typically between two to twelve different owners per property. This translates into each owner spending more time in the property, building an emotional connection, thereby usually taking more care over the appearance of the property, as well as being able to see the enhancement in the value of their share of the property.

Another key differentiating feature is the actual structure of the ownership interest. Where timeshare owners had no control over the property, modern fractional agreements tend to be more akin to a 'home-owners' association' and joint decisions are made by all.

Fractional ownership developers tend to promote the benefits of such ownership to individual investors above SOPU schemes as follows:

- Lower initial acquisition costs – investing in part of a property requires less investment. In addition, the costs of renovating, furnishing and outfitting the property are shared by the owner group.
- Lower operational costs – owning real estate involves ongoing ownership and operating costs such as property tax, insurance, utilities and maintenance, as well as the cost of repairing and replacing the furniture and other household

goods inside. Owning only part of the property means the liability for the fractional owner is lower.

- Reduced operational issues – being part of a multi-ownership resort where a reputable management team is in place running the whole resort means that the individual owners do not have to worry about finding tenants (when they are not using the unit themselves), undertaking repairs and maintenance or dealing with service suppliers. That is all taken care of by the operating company.

Single owner-per-unit ownership (SOPU)

In the last twenty years, there has been an increase in the use of a different type of multi-ownership investors. The difference with the latest schemes is that the whole of each 'unit' (whether a villa, apartment or even a hotel room) are owned by one third-party owner in its entirety, rather than several owners owning part of that property.

This tends to be beneficial to the developer and the resort operator, as there are fewer entities to deal with. If the hotel has 100 units then having 100 (or fewer, if some owners buy multiple properties) is usually preferable to having up to 1,200 fractional owners, or even 5,200 timeshare owners.

The typical reasons for individual investors to purchase part of a multi-owned resort is either as a straightforward investment or for personal use.

The returns involved in certain schemes can be quite attractive, reflecting the relative risk of such an investment. These risks are mitigated by good design, good management and careful choice of the resort, whilst the pricing is usually less responsive to these lowered risks, as the mainstream investment market remains wary of such resorts.

Other purchasers are attracted by the opportunity to buy 'in effect' a holiday home, but one that will usually cover all its ongoing operational costs by letting out the space when the purchaser is not using the property themselves.

SOPU schemes can be either unit-specific (the owner has purchased villa number 87) or non-unit-specific (the owner has purchased 1/100th of all the residential element of the resort), where the property the owner will use depends on the usage allocation system.

Typically, unit-specific schemes appeal more to purchasers who have invested for personal use, as it is easier to become attached to one specific property. In contrast, non-unit-specific schemes tend to be more beneficial for investors, as it is easier to review cashflows and returns when there is no emotional investment in a specific property.

What are the key attractions of multi-ownership resorts?

Resort developers

The main reason resort developers use such schemes is the access to capital it provides. Each 'buyer' effectively helps fund the development of the scheme, and multi-ownership schemes have helped develop resorts that would otherwise not have been able to be funded through traditional methods with bank debt.

In other schemes, this investment effectively replaces the equity contribution required from the resort developer.

The complications involved with creating such schemes that work in the long term are more than balanced out by the additional capital available.

Another attraction to developers is that through the partial sale of the resort they can recoup their initial investment, whilst still retaining ownership of part of the resort. On traditional resorts, it is usual for the developer to have to sell the whole resort to secure an exit.

Individual property investors

Individual investors buy into such schemes for one of two key reasons:

1 investment – earning an income and/or capital growth
2 personal use.

In addition, there are some investors who are interested because of the combination of both the investment returns and the personal use opportunities, but these tend to be less important overall as a specific grouping, as marketing will be concentrated on one or both of the above groups as the most significant market segments.

Owners' use-structure in multi-owned resorts

The usage allocation system for a multi-owned resort decides who can use the property when and, in fractional arrangements involving usage of more than one property, which home an owner will visit. There is no best usage allocation system. Whether a system will work well for a particular fractional property will depend on the location, property characteristics and target market. Usage structures can be fixed, variable or a combination of fixed and variable.

Fixed usage

Fixed usage systems assign each owner specific periods of usage that remain the same each year, for example, the Harpers will have the first two weeks of July every year. More desirable usage (such as high season or school holidays) is typically priced higher reflecting the fact that more buyers want it. The benefits of such a usage system are:

- offers absolute predictability and consistency;
- facilitates relationships among neighbours and allows extended families to more easily coordinate visits;
- allows some owners to save money by buying off-season dates;

- inexpensive and simple to operate (keeping owner dues lower);
- exchanges can provide variation of vacation timing.

Variable usage

Variable usage periods are determined by establishing pre-set groups of days, weeks or months which rotate among the co-owners annually. For example, 'usage package 1' involves the first week of every month while 'usage package 2' involves the second week of every month, and so on. Then the Harpers are allocated usage package 1 the first year, usage package 2 the second year, and so on until they have had all the usage packages, at which time they begin again with usage package 1.

The benefits of such a usage package are:

- lowers cost of high-season usage by allowing more owners to share it;
- eliminates risk that life changes will make assigned period unusable;
- eliminates potential difficulty of finding owners willing to purchase undesirable usage periods.

This type of variable usage assignment allows for unlimited predictability and advance planning and, of course, a co-owner can simply trade with another co-owner if his/her usage for a particular year does not work for his/her schedule.

Alternatively, variable usage periods can be pre-selected annually by the co-owners based on a rotating system of selection priority, or use a more flexible reservation system (although operating a CRS tends to be too complex and expensive for most independent fractional owner groups).

Variable usage: simple reservation structure ('draft system')

In a simple reservation system, each fractional owner reserves all of his/her usage for the following year at a particular time or during a specific window of time. A system of rules determines how the selection process works.

An example of a simple usage reservation system for eight fractional owners would be where, during September of each year, the various owners select their preferred usage for the following year, rotating among themselves as to who chooses first. The owner with first choice in a particular year might choose two weeks, followed by the owner with second choice, and so on. After each owner had selected two weeks, the process repeats in reverse order, then repeats again in the original order. A simple reservation system has the following advantages:

- provides flexibility for owners to adapt to tailor each year's usage without relying on owner exchanges;
- less expensive and simpler to operate than complex usage reservation systems.

Figure 14.2 Fraction development in Kenya

Variable usage: unlimited usage reservation structure

Unlimited usage reservation structures combine a limited amount of advance reservation with unlimited shorter-notice usage. For example, a private residence club with eight fractional owners per home might allow each owner to reserve four weeks of usage per year in advance based on a rotating system of priority (resulting in thirty-two reserved weeks).

When advance reservations close, the twenty-two unreserved weeks become available on a first-come-first-served basis, with restrictions on how far in advance such reservations can be made, and on how many reservations can be held by each owner.

When this type of reservation programme is combined with non-unit-specific usage, owners with the flexibility to vacation at short notice are likely to be able to visit the property far more frequently than with other systems. There are some advantages and disadvantages of unlimited usage reservation systems:

- unlimited usage and last-minute getaways can spur sales;
- systems are expensive to operate (raising owner dues);
- systems are prone to mistakes and breakdowns;
- systems are complex to explain even to discriminating buyers.

Main operational issues from multi-ownership resorts

From the resort owner's (and operator's) perspective, there are a number of key issues that arise from multiple ownership of a resort. These issues need to be

Figure 14.3 SOPU development in Cape Verde

addressed in advance of the individual unit sales if they are not to have an adverse impact on the viability of the scheme.

The main differences in issues arise from the type of owners. People who have bought the unit for private use tend to generate different problems than those who have invested as a simple investment.

Private-use owners – covenants of sale

Private-use owners are investing in the product to use it mainly for themselves or their family. This level of personal use tends to mean that most potential operational issues relate to the use of the property, rather than on investment returns.

If the sales agreement of an individual plot does not lay out enforceable controls over the owner's use of the property, then it is possible the owner can use the property in a way that negatively impacts on the performance of the resort.

These obligations should obviously include the actual use to which the property is put, but also have stipulations as to excessive noise or odours.

In addition, the condition and appearance of individual units needs to be regulated to ensure they look part of the whole resort. If one property is allowed to fall into disrepair, it can lower the guests' perception of the whole resort, thereby impacting trading and value.

In the same way, if one property is decorated in a manner not in keeping with the rest of the resort, it can adversely impact the rest of the resort.

Investment units – covenant of sales

The majority of people who invest in such units do so as an investment and are looking for their property to be let as part of the resort's letting inventory. There are many ways such an agreement can be structured to work, but the general rule is that the owner allows the use of the property by the resort management company for a specified period each year, in return for an income.

Some owners allow the management company to let their property 365 days a year, whilst others reserve use for themselves over set numbers of weeks per year, whilst others still block certain periods off each year where they cannot be let. The structure of such use is a matter between the parties, and none have inherent defects or advantages over and above any other, though there will be tax differences between structures depending on the jurisdiction.

Remuneration for the owners can also be structured many ways. The most usual is a percentage of turnover, or fixed returns. These have the disadvantage of placing financial obligations on the resort owner that may not be able to be met if trading is lower than anticipated. The resort will have certain fixed costs that need to be met, and if the revenue generated is low, then even giving away only a proportion of turnover can mean it struggles to meet its obligations.

It is essential for the long-term future of such a resort that it has the ability to ensure it is all kept in good condition. All trading properties need to invest not only in a repairs and maintenance budget, but also in an FF&E reserve. This will ensure the property is kept in the sort of condition that guests want, and more importantly are prepared to pay for.

It is also vitally important that there is one entity responsible for coordinating the repairs, typically the operating company. However, that entity will not be able to effectively deal with many different owners, so it is usual to have all owners get involved as a single entity (like a condominium owners' agreement), so discussions are between only two parties for the whole resort.

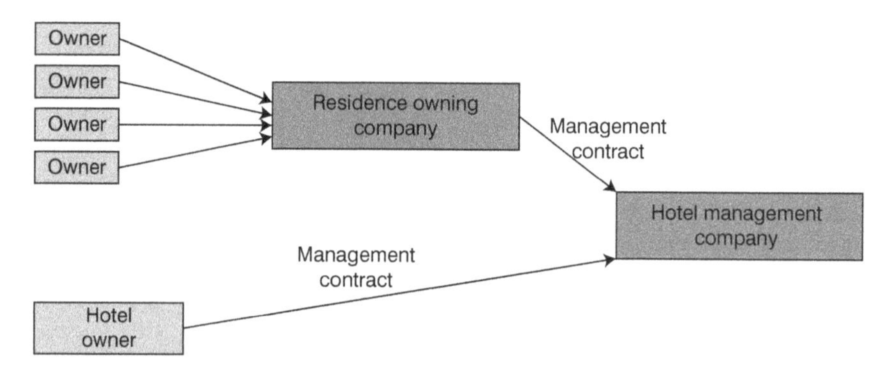

Figure 14.4 Structure outline of typical SOPU scheme

Basic valuation methodology

As with all trading properties, the nature of each business is so varied that direct comparable evidence is difficult to find. It is extremely unlikely that there will be two resorts similar enough to each other to be able to provide accurate comparable advice.

In addition, the variety of potential structures of the multi-ownership profile of resorts is very wide, making the comparable transaction potentially less relevant. Also, the lack of multi-ownership resorts, and the lack of sales on the open market generally, tends to lead to a scarcity of comparable evidence that valuers can rely upon.

As such, the preferred method of valuation tends to be the 'profits method', although it must be stressed that comparable evidence is still important for each part of the valuation under the profits method. Comparable transactions where available will help define yield selection, whilst comparable trading profiles will help determine potential trading profitability for the property.

Example: The Anderson and Broad Resort

In this example, the Anderson and Broad Resort is open and running. It has a letting inventory of 601 rooms, of which 550 are villas and apartments that have been sold off to individual owners. The property has a selection of restaurants and bars, a spa, meeting space and a number of retail units.

The property is managed by an operator on a management contract, paying 4 per cent of turnover as a base fee and 10 per cent of AGOP as an incentive fee.

The owners of each individual apartment are paid 50 per cent of the net revenue generated by their apartments, based on an allocation of all income received from letting units, rather than a detailed list of how often each individual unit was let.

The nature of the property ownership structure means that the capitalisation rate is relatively soft at 12.0 per cent, resulting in a value of €30.325m.

Operational challenges – getting to the bottom line

When carrying out the valuation of such a resort, getting to the sustainable EBITDA of the property is the immediate goal.

It is usually performed through five quite separate exercises:

1. property inspection
2. analysis of the historic trading accounts (last three years)
3. review of the competition
4. interview with the manager
5. independent analysis of everything that has been seen and said throughout the process to determine how the resort is likely to trade in the hands of a reasonable efficient operator (REO).

Table 14.1 Anderson and Broad trading projections

Years	*1*		*2*		*3*		*4*	
	2016	%	*2017*	%	*2018*	%	*2019*	%
Occupancy	52.0%		57.0%		60.0%		62.1%	
No. of rooms	601		601		601		601	
No. of rooms sold	114,070		125,038		131,619		136,226	
ARR (€)	112.00		115.00		118.55		123.85	
Revenue (€m)								
Rooms	12.78	66.4%	14.411	62.0%	15.472	62.0%	16.815	62.0%
Food and beverage	5.49	28.5%	6.740	29.0%	7.237	29.0%	7.865	29.0%
Spa	0.67	3.5%	0.812	3.5%	0.942	3.8%	1.039	3.8%
Revenue other	0.30	1.6%	0.790	3.4%	0.848	3.4%	0.922	3.4%
Total revenue	19.24	100.0%	23.243	97.9%	24.956	98.0%	27.120	98.0%
Operating costs (€m)								
Rooms costs	2.17	17.0%	2.306	16.0%	2.244	14.5%	2.438	14.5%
Food and beverage costs	3.51	64.0%	4.112	61.0%	4.270	59.0%	4.640	59.0%
Spa costs	0.44	65.0%	0.447	55.0%	0.518	55.0%	0.572	55.0%
Other costs	0.05	15.0%	0.119	15.0%	0.127	15.0%	0.138	15.0%
Total operating costs	6.17	32.1%	6.982	30.0%	7.159	28.7%	7.788	28.7%

Deductions (€m)								
A&G	1.73	9.0%	1.859	8.0%	1.947	7.8%	2.115	7.8%
HLP	2.69	14.0%	3.022	13.0%	3.182	12.8%	3.458	12.8%
Marketing	0.67	3.5%	0.814	3.5%	0.873	3.5%	0.949	3.5%
R&M	0.38	2.0%	0.465	2.0%	0.499	2.0%	0.542	2.0%
Total deductions	5.48	28.5%	6.159	26.5%	6.501	26.1%	7.065	26.1%
GOP	7.59	39.4%	10.101	43.5%	11.296	45.3%	12.267	45.2%
Other deductions (€m)								
Management fees	0.77	4.0%	0.930	4.0%	0.998	4.0%	1.085	4.0%
Incentive management fees	0.68	3.5%	0.917	3.9%	1.030	4.1%	1.118	4.1%
Insurance	0.15	0.8%	0.186	0.8%	0.200	0.8%	0.217	0.8%
Rooms profit return to owners	2.72	14.1%	3.587	15.4%	3.929	15.7%	4.130	15.2%
Property tax	0.36	1.9%	0.430	1.9%	0.462	1.9%	0.502	1.9%
FF&E escrow	0.19	1.0%	0.465	2.0%	0.749	3.0%	1.085	4.0%
Total other deductions	4.87	25.3%	6.514	28.0%	7.367	29.5%	8.137	30.0%
NOP (€m)	**2.72**	**14.1%**	**3.587**	**15.4%**	**3.929**	**15.7%**	**4.130**	**15.2%**

Table 14.2 Anderson and Broad Resort income capitalisation method (in €s)

Stabilised EBITDA in present values	4,130,345		
Capitalised at 12.0%	8.333	34,419,542	
			34,419,542
Less income shortfall		2,158,120	
Gross value			32,261,422
Less transaction costs	6.00%	1,935,685	
Net value			30,325,737
Net value – SAY		30,325,000	

Assessing the likely trading profile of the resort will require a complete understanding of the trading environment in which the facility sits. It is difficult to provide a comprehensive check list of things that need to be considered, as multi-ownership resort course tends to be quite different.

However, breaking down the trading profile into sections is usually desirable, looking at the various income streams, followed by the various operational costs, is one of the most sensible approaches as it will usually result in a thorough analysis of the likely vagaries of that specific business.

Income

Typically, a resort will generate income from a number of different sources, and it is essential that the valuer reviews each of these in detail to be able to accurately project the likely income stream for the resort going forward.

These may include the following.

Accommodation

The single largest income generator for most resorts tends to be the accommodation department. As such, it is important that the valuer understands the nature of the accommodation, and how it relates to the potential competition.

Some resorts offer 'all-inclusive packages' where guests pay a single fee to stay at the resort with a number of other facilities, for example, all meals and drinks included within the price. In such circumstances, it is important to take care when 'allocating' a proportion of the income to various departments to ensure that the allocation is fair.

The better quality and more highly desirable the resort, the higher the number of visitors and the higher the accommodation charges tend to be. As is the case for hotels, the best management team will juggle visitor numbers with an average rate to generate the highest possible RevPAR. The key difference between resorts and hotels is that many resorts have a high level of incremental spend, with guests paying extra for massages and other similar services, so there tends to more emphasis on TRevPAR than there typically is in more traditional hotels.

Food and beverage

Typically, there will be more than one F&B outlet, possibly a buffet restaurant, the main bar, an à la carte restaurant, pool bar and possibly a beach restaurant, all in addition to room service.

In resorts where owners have units for their own personal use, there will be a different F&B spend profile to those where the units are purchased to be placed into the letting inventory. However, even units not included within the letting inventory can generate significant F&B income if handled properly by the operator, and it is important to understand how the dynamics of the resort works in this regard.

It is also important that the valuer is able to determine the operational efficiencies of each outlet, along with its suitability for the current (and future) market, to allow an assessment of likely future trading to be accurately made.

Retail

Resort shops can generate reasonable income streams, and vary on different properties from supermarkets (very important where a large number of the units are not contained within the letting inventory) to souvenir shops and convenience stores.

Spa

Most resorts now tend to have spa facilities of some sort, running from full spa services to a couple of treatment rooms attached to a more basic gym. The quality of the facilities and staff, along with the promotion of such services and the level of immediate competition, will all determine how much income is generated from this department.

Sports

Some resorts will charge for use of facilities, whether a round of golf, scuba-diving lessons, hiring the tennis courts or using the water sports facilities. The facilities, pricing and level of demand will all impact on the revenue generated from such sources.

Other income

There is usually a separate 'other income' line in the accounts to cover all income that is not included in specific individual lines. It might include items such as rental income, advertising or private hire, for example.

Costs

Costs are usually broken down into the various departments outlined in USALI. These include the direct cost of sales by department, sales and marketing, laundry, administration and general, repair and maintenance, water charges, insurance and taxes, among others. Some resorts may provide staffing costs as a separate department, rather than allocating the costs to the correct department, as is standard practice. It is important for the valuer when assessing various cost lines with comparable trading entities to ensure that staff costs have been handled in the same way in both instances.

Inspection

A typical multi-owned resort inspection will follow the general course of a hotel inspection. Initially the valuer will request the information required, ahead of the inspection. This will include all, or some, of the following:

Legal information

- title documentation or reports on title
- ground leases or property leases
- schedule of all owners and details of their various ownership interests
- copies of the ownership agreements
- subleases
- occupational licences or concessions
- planning consents for the site and each building on the property
- any restrictive covenants and other use prohibitions
- details of any unused planning consents
- details of any unfulfilled planning conditions
- copies of the operating licences
- agreements in place with any occupiers or other professionals, etc.
- details of any 'unsold' units
- details of sale prices and dates, including any 'resales'.

Property information

- area of the overall site
- site plan showing ownership interests
- site plans (with sizes) for each building
- building plans – floor by floor
- breakdown of accommodation facilities (type, number, size)
- breakdown of other facilities including F&B outlets
- copies of any current or historic valuation reports
- copies of any current or historic condition surveys, asbestos surveys, etc.
- details of historic capital expenditure on the properties (last three years)
- details of any proposed capital expenditure (this year and next year's budget).

Operational information

- historic profit and loss accounts (three years minimum); this will be the full P&L accounts, not just the summary sheet
- example of typical terms and conditions of the ownership agreements
- management statistics, including occupancy, ADR and RevPAR
- breakdown of business by packages sold (all-inclusive, half board, full board, room only, etc.)
- statutory accounts
- full profit and loss budget for the current year and for the next three years (if possible)
- details of market segmentation and geographical analysis of the guest profile
- average lengths of stay of guests
- staff details – numbers of staff and contracts of employment, including pension details, length of service and skills training for each member of staff
- sales and marketing plans; this year and next year's budget
- details of F&B revenues; where it comes from, covers, average spends, etc.
- breakdown of F&B revenue comparing resident and non-resident spends
- details of the local markets; SWOT analysis of the local competitors
- details of local trading; how does the property compete with its competitive set?
- details of leased operational equipment.

Other

- details of any outstanding complaints, legal claims, etc.
- details of the tax positioning of the property (outstanding tax bills, capital allowances, etc.).

The resort will be inspected, usually in the presence of the general manager. The valuer will be looking at the resort facilities to see that it is suitable for use and how appealing it will be to potential guests. In the case where parts of the residential component remain unsold, they will be reviewing how appealing the residential portion will be to potential purchasers.

This will differ when looking at the benefits of the resort to customers, and the benefits of the residential component to potential investors. However, in essence, the valuer will be looking for factors that will impact on trading, as well as looking for factors that will enhance or detract from the investment opportunity.

In effect, the purpose of the inspection is to see all matters that may impact on revenue generation or have cost implications, so an accurate assessment of future trading can be made. The condition of everything will also be assessed to ensure that any future capital expenditure requirements can be accurately gauged.

The quality of the resort is then assessed in relation to the relevant market, so an accurate assessment of the appropriate yield for the property can be made to finally determine value.

Interview with the general manager

After the inspection of the property, the trading history, past, present and future, of the resort will be discussed with the management.

Typically, a detailed discussion will review the operational issues relating to the resort in terms of revenue generation, followed by anticipated costs resulting in an understanding of the future trading profile and a review of the proposed capital expenditure plans.

The local and international markets can be key to revenue generation for the property and new supply, even in different countries can potentially impact on future trading. As such a detailed discussion with the general manager about the various strengths and weaknesses of the resort in relation to the existing (and proposed) market is desirable to allow the valuer to understand the local market dynamics as seen from the perspective of the current manager.

The income generated by each segment, including accommodation, F&B, spa and other revenue will be discussed in light of past trading so an understanding of the projections can be fully made.

The same process will occur for each revenue line, followed by each cost line to ensure the trading projections are typical of the local market, and achievable.

The valuer will ask to review any licences that are required for operation, discuss the ownership of the property, ownership of the equipment and details of any franchises or concessions.

Discussion will also revolve around the various different owners and how well the various interests combine, as well as common complaints or recurring problems.

Example: Haynes and Greenidge Hotel, Resort and Spa

In this example, the resort is undeveloped. The same operating strategy is anticipated as for the Anderson and Broad Resort except this time it is a proposed development that has not been started.

The management contract is in place, firm building costs on a D&B contract have been negotiated and a number of buyers have already purchased properties 'off-plan'.

The trading projections for the resort are shown in Table 14.3.

Five hundred-and-fifty of the villas will be sold off in the first four years, at an average net present value of €400,000 per unit. In the first year 100 are sold, in the second year a further 225 are anticipated to be sold, with 140 in year 3 and the remaining 85 in the fourth year.

The property will take two years to build, with €95 m spent in the first year and the remaining €60 m in the second year (including all costs and finance).

Taking into account the risks associated with the project, the discount rate is 17.5 per cent, and transaction costs are estimated to be 2 per cent, resulting in a site value of €35.3 m, which equates to approximately 12.7 per cent of the GDV.

Branded residences

Daniel von Barloewen – director, head of Savills International Development Consultancy

Over the past two decades, the hospitality industry has diversified through the creation of budget brands, and, subsequently, serviced apartments. A further product type (present since the mid-twentieth century but of rapidly increasing popularity in the past decade) is the branded residence product. While not strictly a hotel product, it is intertwined with hotel brands.

Typically, branded residences are the result of a partnership between hoteliers and developers. Under an agreement, the hotel operator brands and manages the residential real estate component of a scheme that is built by the developer. Individual residential units are then sold to private individuals or entities.

The majority of branded residences ownership frameworks and subsequent rental pool structures fall into one of these three categories:

1 The development consists of a hotel component and a branded component of fully furnished residential units for sale, with a 'mandatory' rental pool. The branded residences will be managed by the operator as serviced apartments in a 'hotel + serviced apartments' structure.
2 The development consists of branded, fully furnished residential units. Part of these will be sold to private investors with a mandatory rental pool requirement. The other part will form a guaranteed number of units for the operator's management, regardless of the rental pool. Sold branded residences will be managed by the operator as serviced apartments, to the extent, and as long as, these are kept in the rental pool in a 'serviced apartments' structure.
3 Residences are branded and sold to owners but with no mandatory rental pool, programme or agreement (some operators do not currently have a rental programme at any of their residences). This is the structure mostly in place in both New York and London.

Branded residences can be found in both urban and resort locations. In urban locations, the branded residences are usually apartments located in the upper floors of a hotel complex. In resort locations, residences can comprise a mix of apartments, penthouses, townhouses or villas in proximity to the branding hotel, depending on land availability and location.

Table 14.3 Haynes and Greenidge Hotel, Resort and Spa trading projections

Years	*1*		*2*		*3*		*4*		*5*	
	2016	%	*2017*	%	*2018*	%	*2019*	%	*2020*	%
Occupancy	52.0%		57.0%		60.0%		62.1%		62.1%	
No. of rooms	601		601		601		601		601	
No. of rooms sold	114,070		125,038		131,619		136,226		136,226	
ARR (€)	112.00		117.00		119.45		125.38		129.75	
Revenue rooms	12.776	66.4%	14.634	62.0%	15.713	62.0%	17.076	62.0%	17.674	62.0%
Revenue food and beverage	5.490	28.5%	6.845	29.0%	7.350	29.0%	7.987	29.0%	8.267	29.0%
Spa	0.671	3.5%	0.812	3.4%	0.942	3.7%	1.039	3.8%	1.076	3.8%
Revenue other	0.300	1.6%	0.803	3.4%	0.862	3.4%	0.936	3.4%	0.969	3.4%
Total revenue	19.237	100.0%	23.604	97.8%	25.343	98.0%	27.542	98.0%	28.506	98.0%
Operating costs (€ '000s)										
Rooms costs	2.172	17.0%	2.342	16.0%	2.278	14.5%	2.476	14.5%	2.563	14.5%
Food and beverage costs	3.514	64.0%	4.176	61.0%	4.336	59.0%	4.712	59.0%	4.877	59.0%
Spa costs	0.436	65.0%	0.447	55.0%	0.518	55.0%	0.572	55.0%	0.592	55.0%
Other costs	0.045	15.0%	0.120	15.0%	0.129	15.0%	0.140	15.0%	0.145	15.0%
Total operating costs	6.167	32.1%	7.084	30.0%	7.262	28.7%	7.900	28.7%	8.177	28.7%

Deductions (€ '000s)										
A&G	1.731	9.0%	1.888	8.0%	1.977	7.8%	2.148	7.8%	2.223	7.8%
HLP	2.693	14.0%	3.069	13.0%	3.231	12.8%	3.512	12.8%	3.634	12.8%
Marketing	0.673	3.5%	0.826	3.5%	0.887	3.5%	0.964	3.5%	0.998	3.5%
R&M	0.385	2.0%	0.472	2.0%	0.507	2.0%	0.551	2.0%	0.570	2.0%
Total deductions	5.483	28.5%	6.255	26.5%	6.602	26.1%	7.175	26.1%	7.426	26.1%
GOP	7.588	39.4%	10.265	43.5%	11.480	45.3%	12.467	45.3%	12.903	45.3%
Other deductions (€ '000s)										
Management fees	0.769	4.0%	0.944	4.0%	1.014	4.0%	1.102	4.0%	1.140	4.0%
Incentive management fees	0.682	3.5%	0.932	3.9%	1.047	4.1%	1.137	4.1%	1.176	4.1%
Insurance	0.154	0.8%	0.189	0.8%	0.203	0.8%	0.220	0.8%	0.228	0.8%
Rooms profit return to owners	2.717	14.1%	3.646	15.4%	3.994	15.8%	4.198	15.2%	4.345	15.2%
Property tax	0.356	1.9%	0.437	1.9%	0.469	1.9%	0.510	1.9%	0.527	1.9%
FF&E escrow	0.192	1.0%	0.472	2.0%	0.760	3.0%	1.102	4.0%	1.140	4.0%
Total other deductions	4.871	25.3%	6.619	28.0%	7.486	29.5%	8.268	30.0%	8.558	30.0%
NOP (€ '000s)	**2.717**	**14.1%**	**3.646**	**15.4%**	**3.994**	**15.8%**	**4.198**	**15.2%**	**4.345**	**15.2%**

Table 14.4 Haynes and Greenidge Hotel, Resort and Spa income stream

	Hotel income €	*Apartment sales* €	*Costs* €
Year 1	–	40,000,000	95,000,000
Year 2	–	85,500,000	60,000,000
Year 3	2,717,121	50,540,000	
Year 4	3,645,553	28,843,900	
Year 5	3,993,712	–	
Year 6	4,198,490	–	
Year 7	4,345,437	–	
Year 8	4,497,528	–	
Year 9	4,654,941	–	
Year 10	4,817,864	–	

Table 14.5 Haynes and Greenidge Hotel, Resort and Spa discounted cashflow (in €s)

NPV net base cashflow		
Discounted at 17.50%		36,011
Additional income		
Discounted at 21.50%		0
Gross value		36,011
Acquisition/sale fees	2.00%	706,078
Net value		35,300,000

15 Gym values and how they are determined

Introduction

This chapter will outline the different types of gyms, how they work as businesses, what customers are looking for when selecting a gym and how they are valued. It will also outline how they work effectively as part of a hotel or resort, and areas where potential problems can be experienced.

According to IHRSA, the market for gyms is growing significantly on a global basis. Over the last five years the number of gyms offering membership rose 40 per cent to 180,000, while the number of gym members rose 21 per cent to over 144.7 million members. If gyms within hotels are included, the total count is estimated at closer to 500,000 units. Problems with obesity, changing work profiles and leisure activities have led to a boom in gym membership in unprecedented numbers as the population try to stay healthy.

The terms fitness centre, gym and health club are often used interchangeably as a lot of the services they provide are similar. However, there are differences and these can impact on the likely trading profile of the business, and therefore on the value of the asset.

The main differences between the key categories are outlined briefly below, but throughout this chapter, the term gym is used to cover all types of business.

Different types of gym

Fitness centre

A fitness centre usually accommodates a variety of different types of exercise and sport facilities and is often aimed at members as well as day visitors. Some fitness centres are focused towards one specific activity or discipline. There are Pilates centres, which focus on conditioning the mind and the body with specially designed equipment and a series of exercises to improve mental capacity, control, balance, flexibility, endurance, breathing and general fitness. Aerobic centres tend to focus on cardiovascular health with the use of equipment, classes and workouts. Yoga centres (of many different kinds and styles) all provide exercises for a healthy body and mind. Dance centres offer classes to improve posture, strength, flexibility, balance, coordination and breathing control.

Fitness centres usually occupy large spaces although there are some smaller units. Size tends to depend on the amenities and activities they offer. Many have a combination of indoor and outdoor sports facilities such as swimming pools, running tracks, rock-climbing walls, sports playing fields, racquetball courts, rooms for martial arts, yoga, Pilates, spinning, weight training as classes or for individual use, and regular gym equipment such as treadmills, elliptical and cycling machines. They may also include saunas, hot tubs, steam rooms, tanning beds, warm-up and cool rooms, showers, locker rooms, childcare services, juice bars, snack bars and retail sports stores.

Health clubs

Health clubs are similar to fitness centres but tend to have a greater emphasis on the overall 'wellness', incorporating mental health, life balance, nutrition and education. They typically provide fitness classes for groups and individuals, a variety of equipment, trained and certified staff. Facilities may include steam rooms, saunas, locker rooms, showers, pools, therapy clinics, spaces for team sports like volleyball and basketball, juice bars, restaurants, television, music and just about everything you will find in a modern gym or fitness centre. It all depends on the size and maybe the location of the facility.

Health clubs also attract a variety of clientele because they provide an environment of comfort. Typically, the larger the health club facility, the more expensive it is to become a member.

Gyms

Historically the term gym meant a weight training gym whereas now it is an all-encompassing term covering a myriad of different designs and business models. Typically, today's gyms offer a variety of indoor activities, amenities and events, and just like fitness centres, they come in many sizes and types.

Gyms typically offer classes in various workout disciplines and sporting activities with specifically designated areas for strength work, cardiovascular training and classes. They may provide experienced, certified personal trainers and staff, and will sometimes have facilities such as swimming pools, saunas, steam rooms, basketball courts, tanning facilities and childcare services.

Typically, the cardiovascular equipment will have televisions, audio equipment or virtual reality devices attached to them.

High-end gyms go all out, providing the services of fitness centres, gyms and health clubs combined and more. For instance, some gyms offer beauty products in their locker rooms, facials, eyelash extensions, massages and clothing boutiques. Others have live DJs and sponsor special events such as 'block parties', while others even provide laundry services to their clients.

There are many different categories of gyms, all aiming at different market segments. Typically, the various types of gym aim at a combination of market segments, with some segments requiring different offerings and services. The various core segments include the following:

- deconditioned market
- family market
- corporate market
- children's market
- seniors' market
- women-only
- minority market
- Generation X market (people born between 1960s and 1980s)
- Generation Y market (also known as the millennials, those born after the 1990s)
- rehabilitation market
- quick-fix market.

The main types of gyms that are commonly found can be categorised as follows:

Weight training gym

Historically, most gyms were primarily focused on weight training, whether concentrating on resistance machines or through the use of free weights. Although many gyms have started to expand the services offered, including cardiovascular workouts and fitness classes, some gyms still focus purely on weight training or lifting.

24-hour gyms

Twenty-four-hour gyms are open at all hours, and are popular with people who tend to have non-traditional work schedules. They can sometimes be more expensive, and although the equipment is available twent-four hours a day, many of the classes are only available during traditional hours.

Women-only gyms

Gym operators realised that sometimes women may feel intimidated by men at a gym because they can feel like they are being watched, ridiculed or judged by men, especially in the weight training area. Women-only gyms were developed to take away this barrier to a very important and potentially lucrative market.

Women-only gyms represent one of the larger growth sectors in the gym market, fulfilling a need that had previously been mostly un-met. Usually, the exercise programmes are customised specifically for women.

Budget gyms

The facilities in a 'no frills gym' are limited to reduce the cost of use. For example, in many budget gyms there are no water fountains, towels or locks for lockers, and sometimes these gyms aren't even manned. Most of these gyms are not open as early or as late as many of the more full-service gyms, but they usually cost significantly less money to use. The reduced cost is usually due to less space utilised, less sophisticated equipment, lower labour costs and lack of luxury services like swimming pools, space, day-care, personal trainers or group fitness classes.

Specialty gyms

Some gyms focus on certain elements of exercise rather than the entire spectrum, for example, yoga gyms, aerobics gyms or kickboxing gyms. The degree of specialisation usually results in a high level of expertise that appeals to a customer base looking for that discipline.

Hotel gyms

Most hotels now offer a gym, though sometimes it can be as little as two bedrooms knocked through offering six or eight pieces of equipment in an unmanned area. These 'fitness rooms' are offered as an incentive enticing guests to stay at the hotel, rather than to generate direct revenue in any way.

Other hotels have extensive fitness facilities that attract significant external membership, and are run on the same basis as most stand-alone gym businesses.

Company gyms

A large number of companies operate in-house gyms as a way to enhance staff morale, differentiate the level of staff care provided by the company with their competitors, as well as to enhance the well-being of employees. They tend not to take external membership and the cost for staff is subsidised, so typical operating margins are below industry averages.

Sports clubs

Gyms that also cater to wider sports tend to be referred to as sports clubs. They tend to attract customers who are keen to play sports such as racquetball, tennis, swimming, football or other sports. Typically sports clubs will be larger gyms with extensive outdoor facilities, and usually have greater membership numbers.

Circuit gyms

A circuit gym is set up to take the guesswork out of which machine the customer should use and the order in which they should use them. Exercises done on the right machine and at the right interval make workouts quick and precise and can result in faster workout times, which can be sometimes important to customers.

Gyms for the overweight

Some gyms specialise in the 'deconditioned' market. Quite often people trying to lose weight don't feel comfortable when surrounded by 'super-fit' people, and a market has grown up providing a 'non-threatening' environment for this market group.

Figure 15.1 Hotel gym

Figure 15.2 Wet-leisure facilities

Typical classes

One of the keys to a successful gym is having a selection of classes that appeal to the members of the gym, and keeping them as loyal customers. Having a well-balanced programme is one of the most effective ways to reduce the attrition rate (the percentage of members terminating their memberships) at a club. Some of the more popular classes on offer are the following:

- aerobics – a vigorous exercise class usually undertaken to music;
- aikido – a Japanese martial art that uses techniques such as locks and throws, and focuses on using the opponent's own energy against himself;
- aqua-fit – aerobic exercises performed in a swimming pool where the water provides support and resistance to increase stamina, and stretch and strengthen muscles. It is particularly popular with older customers;
- ashtanga yoga – a type of yoga class that helps you to work on deeper stretching, focusing on correct alignment and pose positioning;
- belly dancing – exercise based on the traditional art of belly dancing, originating in Asia;
- body attack – exercise class which involves intense cardiovascular workout for weight loss and body toning;
- body balance – fitness class with a mix of yoga, tai chi and Pilates techniques that help build strength and flexibility;
- BodyPump – a group fitness class that uses free weights;
- boxercise – combination of an aerobics and boxing workout;
- boot camp/cross training – circuit training/cross training has many forms but primarily involves moving from one station to another in set periods of time, each interval varies from 1–3 minutes alternating between strength and cardi; cardio work can include step, hi/lo aerobics or cardio kickbox, while strength work can include physioballs, bands, mat exercises and body weight exercise;
- cardio kickbox/turbo kick – this format combines various forms of martial arts and self-defence moves with aerobics, producing a workout for the whole body;
- circuits – a series of exercises that promotes the body's flexibility, strength and endurance; each station has a different exercise and each participant moves around the circuit, visiting each station;
- dancercise – dance movements used as an aerobic exercise;
- hatha yoga – this type of yoga combines postures and breathing techniques to help build flexibility and relax the mind;
- high-impact/high-energy aerobics – this form of aerobics involves jumping, jogging and hopping movements where both feet lose contact with the ground;
- hot yoga – where yoga is practised in a room heated to 45°C, also sometimes referred to as bikram yoga;
- kickboxing – a martial arts and combat sport that originates from karate, muay Thai and western boxing, which uses kicking and punching to help general fitness and self-defence;

- low-impact aerobics – form of aerobics with side-to-side marching or gliding movements which spare the body from excessive stress and possible injuries;
- meridian stretching – stretching exercises designed to encourage physical and mental flexibility, for the body and mind to perform at their peak; combines exercises, yoga and traditional Chinese medicine;
- muay Thai – Thai boxing that involves the use of the upper and lower bodies. A cardiovascular and fat-burning exercise that helps relieve stress;
- physioball – this non-aerobic class uses a large air-filled ball for strength, flexibility and balanced exercises;
- Pilates – a combination of stretching and strengthening exercises, involving intense mental concentration on the movements;
- spinning – intense group aerobic exercising using a stationary bike at various paces and led by an instructor;
- step aerobics – aerobic sessions done with a small platform for stepping up and down;
- tai chi – developed in China, it is a form of martial arts that is learned for defence and health reasons. A form of moderate slow-motion exercise, a type of moving meditation, and a system of martial art;
- vinyasa yoga (or 'flow yoga') – this is a style of yoga in which the poses are held for only a short while so they move from one pose to the next, building strength and fluidity of motion;
- yogalates – a blend of two popular types of mind-body connected exercise: yoga and Pilates;
- zumba – aerobic fitness programme inspired by Latin American dance, performed primarily to Latin American dance music.

Design parameters within a functioning gym

The target market of the gym along with the user demographics will impact upon the design requirements and size of the gym. It is vital that the valuer identifies the target market at an early stage so they can assess the suitability of the physical product to meeting the operating needs of the business.

Ideally, the overall space will be flexible and able to accommodate new classes, programmes or trends that may become popular in the future, allowing for a smooth evolution of the business when needed.

In the following section, design parameters are discussed, detailing how they impact on the business. Many gyms will not meet all these parameters, and if that is the case it is important that the valuer can recognise any shortfalls and can reflect them in the trading projections and capitalisation rates applied to the valuation.

As previously mentioned, there are a wide range of potential disciplines, all requiring different layouts and facilities, and if the area is unsuitable for its discipline then it is likely to impact on long-term trading patterns.

A typical gym will usually include a fitness gym area (containing strength and cardiovascular equipment), studio spaces, reception area, café/bar, changing rooms (including showers and toilets) and other sundry facilities.

It is important that the valuer reviews the suitability of each element from an operational perspective to assess the potential impacts on the trading profile of the business.

Reception

The first point of contact at a gym should always be the reception desk. Fitness facilities forming part of a hotel or resort will still require a separate reception point with its own access control system. The fitness entrance should be clearly visible and welcoming.

Access to the gym is usually controlled from the reception desk by an access system, whether turnstiles or gates. The design should have the ability to cope with peak throughput, making provision for large numbers of users, avoiding the need for them to queue. The reception desk is also the primary point for obtaining information, booking and purchasing goods, and as such should be suitable for use, as well as aesthetically pleasing. It may also be a point for issuing and collecting towels. Therefore, the functionality of design is important.

Ideally, the reception desk should have a lowered section allowing access for children and wheelchair users. The staff side of the reception desk should have direct access to the office area.

Office

A secure office will be needed to accommodate the day-to-day administration and general housekeeping duties. Ideally, there should be a physical link with the reception area, although a visual link is acceptable if the physical link is not possible. The office should be large enough for back of house functions, and may incorporate the customer sales area.

Customer sales and marketing area

Most branded operators often require a dedicated private area away from reception to provide promotional marketing and hospitality without distracting staff operating the main reception desk.

This may be a quiet area in the main foyer area, or preferably a separate room. New customers may be given details on the centre, prices, facilities and a tour. If adequate provision is not available for this feature, it could impact adversely on this element of the business, potentially decreasing value.

Changing rooms and toilets (including accessible facilities)

The quality and suitability of changing rooms is one of the key potential areas of complaints with members, and is possibly the single most quoted reason when members leave a gym. As such, they are very important to the trading potential of the business. Changing room capacities and sizes should meet (or exceed) the likely normal maximum occupancy level and patterns of use, otherwise operational issues (and complaints) will occur.

Ideally, changing facilities should be single-sex facilities with a 'buffer changing area' that can be used by either sex for peak demand or for large groups. In addition, it is preferable that the gym provides individual unisex accessible changing rooms in addition to providing full access to the larger single-sex changing rooms.

Where the centre also includes a wet facility, the changing rooms should be designed to separate wet and dry foot traffic.

Changing rooms should typically consist of changing areas, toilets, shower areas, lockers and a vanity area with mirrors and hair dryers.

Changing areas vary from open bench seating through to individual changing cubicles, and the customer base will dictate which is preferable.

Each type of gym will require an individual assessment of capacity and layout, however, as a guide, the following assumptions are a rough rule of thumb of what may be required.

Fitness gym

Where the fitness area is relatively small, ideally, there should be one changing space provided for each item of equipment. For larger centres, a lower ratio of changing spaces and lockers are required (usually discounted by 25 per cent to 35 per cent on the number of work stations) as gym usage is individual and users arrive and leave at different times.

There should be approximately one shower for every six changing spaces in the fitness gym.

Studio changing requirements

In addition to fitness gym members, the changing rooms will also need to be able to cater for the demand of the participants of studio classes, particularly at peak periods as classes start and end.

Typically, there should be one changing space for each 5m^2 of studio floor area.

In addition, there should be one and a half lockers for each person using the studio(s) over a one-hour period.

There should be one shower for every six changing spaces within the studio.

Fitness gym

The overall fitness gym area will depend upon the anticipated number of users and mix of equipment. The minimum required space is 25m^2, although the majority of gyms occupy an area of 100–200m^2 to ensure a range of options are given to users.

The height of the ceiling is important. The optimum ceiling height should be between 3.5–4m from finished floor level and should not be lower than 2.7m, as this would limit the use of some exercise equipment. Lower ceiling heights also lead to a sense of claustrophobia amongst some customers, as well as making it more difficult to supress odours and regulate room temperature.

The shape of the fitness gym area will inevitably be defined by the overall design of the building, but ideally it should be broadly rectangular with a length to width ratio below 3:1. Adequate space in the gym is important to ensure the required range of equipment and facilities is comfortably accommodated.

One common reason for customer dissatisfaction is too little space between equipment. A well-spaced gym will typically have a floor area of $5m^2$ per piece of equipment. This includes an allowance for circulation space around the equipment.

The equipment mix will depend upon the target market. For general use, the split ratio of cardiovascular (CV) equipment to resistance equipment should be between 40 per cent to 60 per cent, however this will depend on local need and demand.

Most gyms (except budget gyms) will also provide a chilled drinking water fountain and paper towel dispenser (for wiping down machines after use).

A fitness gym is usually separated into individual zones, determined by equipment or exercise type.

Stretch area (warm up and cooling down)

The stretch area is usually the first and last area to be used, for warming up and stretching limbs before moving on to other equipment, and cooling down afterwards. Typically, it is located near to the fitness gym entrance to promote its use. The area will ideally have padded floor matting and wall-mounted mirrors. A wall bar is usually provided at a height approximately 1.2m above the floor for support. There is normally an un-mirrored wall area provided for stretching against.

Cardiovascular area

The CV area will ideally be on a single level and contain fitness machines with integral visual displays and audio output (normally headphones). Equipment can usually be linked to a centralised fitness monitoring programme requiring data connections.

CV equipment will normally include a number of machines, each designed to provide a different form of exercise, arranged in combination. The equipment may include the following:

- tread or running machines
- upper-body ergometers, fitness and exercise machines
- cross trainers
- bicycles
- step machines
- rowing machines.

The average user may spend between twenty and thirty minutes on each piece of equipment. CV areas tend to be planned so that the CV equipment is arranged in multiple tiered rows facing one direction, with the lowest equipment at the front, grouped in front of an audio/visual (AV) system. The AV system will normally include a number of large plasma flat screen televisions, wall or gantry mounted, with each screen offering a different entertainment channel. Switched headphone sockets on the exercise equipment allow each user to select the sound channel relevant to the programme/screen being watched.

Resistance and/or free weights area

Between eight and ten pieces of different equipment usually provides an adequate range of exercises for most users. The average user may spend between three and six minutes on each piece of equipment. The space needed for each exercise machine varies considerably and careful assessment of the equipment is important, as changing the equipment could lead to a visible improvement in the layout, and potentially lead to increased revenue potential for the business.

Mirrors should be provided for users to check their positioning while using the equipment.

The resistance area equipment should be arranged logically and typically is zoned by exercise type, allowing users to move through the station in a logical manner.

Typical machines may include the following:

- high/low pulley
- chest press
- shoulder press
- seated row.

In addition, most gyms will also provide a range of small loose equipment (e.g. low weight dumbbells, dynabands, etc.)

Floor finishes need to be slip, stain and static resistant, and fit for purpose for a gym environment. The floor finish should contrast in colour to the equipment to ensure that equipment is less likely to be a trip hazard.

Assessment rooms

Assessment rooms are usually provided to allow consultations with personal trainers to be undertaken confidentially. The room may accommodate the following:

- desk and comfortable chairs for trainer and trainee
- telecom/computer equipment
- heart rate/pulse monitoring equipment
- weighing scales and height gauge
- secure storage for valuables and records
- information board.

If the assessment room is also used for first-aid treatment, a hand washbasin, a stretcher bed and basic life-saving equipment would typically be required.

Studio(s)

The number of studios and the size of each studio required will be determined by:

- number of simultaneous classes
- type and range of programmes
- frequency and duration of each class
- number of attendees for each class.

The studio ideally should be square, or rectangular with a length to width ratio of approximately 3:2. Instructors generally stand facing the users on the long side of the studio. Columns, projections and splayed walls all adversely impact on the amenity value of the studio. Optimum sizes will vary, dependent upon the use of the studio.

When looking at the suitability of a studio space, the valuer will need to bear in mind:

- the current market
- changing trends over time
- the range of activities to be catered for.

Most classes such as 'Step' or 'Legs, Bums and Tums' require the simple addition of lightweight equipment and can be undertaken in a general purpose studio. However, a number of classes require specialist rooms, some of which are expensive to provide and may not prove popular in the longer term.

Each geographic region has different preferences for classes and it is not possible here to outline them all.

A multipurpose studio of either 15 × 12m or 15 × 15m is generally an optimal size. Although studios larger than those shown in Table 15.1 could be used, consideration should be given to the maximum size that will suit a single class.

The following classes requiring specialised spaces have proved to be popular and durable around many parts of the world.

Spinning

Spinning classes generate substantial noise and activity disruptive to other users. Spinning should therefore ideally be separated from the main gym and held in a separate studio. Dedicated spinning studios are preferred as the equipment can be bulky and awkward to store. These can be smaller than exercise studios as there is no extended movement, other than for getting on, off and general circulation around the equipment.

Audio-visual equipment using screens and soundtracks showing a moving landscape over which the classes visualise cycling are offered as a more interactive

Table 15.1 Multipurpose studios

	Typical dimensions
Movement (small groups)	12 m × 9 m
Movement (average groups)	12–15 m × 12 m
Movement (large groups)	21–24 m × 12m

experience. Spinning classes can vary between small and large groups of twenty or more cyclists.

Spinning studios may require specialist lighting, projection or plasma screen TVs and audio systems.

Pilates

A low-impact stretching and conditioning exercise that builds core strength, improves posture and flexibility through small repetitive movements. Pilates can be offered as a mat-based session, but more intensive classes involve a range of specific Pilates equipment. The apparatus requires a permanent studio as Pilates requires concentration, low noise levels and the equipment is too large to be moved regularly. Pilates is practised by men, women and children, from a recreational level through to professional athletes.

Yoga

A studio, mat-based class. Yoga is a tranquil exercise, based on body positions, controlled breathing and meditation. Studios require good acoustic separation to allow for quiet concentration. Yoga and other meditation-based exercise classes may also require privacy and low levels of illumination or specialist lighting. Blinds to all glazing should be considered.

Martial arts studios

A wide variety of martial arts are practised by all ages, and prove popular with many gym members.

Kickbox/boxing-aerobics

These classes mix traditional aerobic exercise with boxing and kickboxing techniques, using gloves, pads and punch bags. They provide high-powered, high-impact workouts.

Dance aerobics

There are many variations of the idea of mixing aerobic exercise and dance, from belly dancing to ballet. The aim is to allow for exercise and fun, while

Table 15.2 Martial arts studios

Type	*Mat size*	*Safety zone*	*Total space required*
Aikido	9 m × 9 m	1m margin, with 2m on one side	11 m × 12 m 3 m high
Chinese martial arts Wu Shu set and traditional routines	14 m × 8 m for routine exercises		14 m × 8 m 3 m high
Judo	6 m × 6 m (junior) 8 m × 8 m (senior)	3 m on all sides 3 m on all sides	10–12 m × 10–12 m 4 m high 12 –14 m × 12 –14 m 4 m high
Jujitsu	10 m × 10 m combat area 16 m × 16 m total mat area	1.2–2 m on all sides	18.4–20 m × 18.4–20 m 3 m high
Karate	10 m × 10 m	1 m margin, with 2 m on one side	12 m × 13 m 3.5 m high
Kendo	9 × 9 m–11 × 11 m (no matting)	1.5–2 m margin, with 2 m on one side	16 m × 17 m 4.5 m high
Taekwondo	8 m × 8 m	2 m on each side	10 m × 10 m 3.5 m high
Tang Soo Do	8 m × 8 m combat area 10 m × 10 m total mat area	3–4 m on each side	14–16 m × 14–16 m 3.5 m high

learning basic dance steps. These classes need additional space compared with usual classes, for the extended movement. In addition to the requirements for a standard studio, a dance studio usually requires:

- wall mirrors and a fixed barre for balance exercises;
- additional lose mobile training barres, with adjustable heights to cater for younger dancers;
- a piano or other musical equipment;
- adequate secure storage for lose/mobile equipment and musical equipment.

Food and beverage

If the gym forms part of a hotel or resort it will rarely have its own F&B outlet. However, where the fitness facility is stand-alone unit, a café and/or lounge area is usually provided, ideally located close to the reception foyer. This area often forms the social hub of the facility, as well as providing an area for relaxation and refreshment before or after using the facilities. The size of any such outlet should

Table 15.3 Dance studios

	Typical dimensions
Small (2–15 people)	9 m × 9 m
Standard (30–35 people)	12–15 m × 12 m
Large	15 m × 17m

be determined to suit the number of users and the proposed menu. There should also be vending machines for times when the café is closed.

A kitchen and servery counter may be a requirement depending upon the size of the facility. The kitchen and servery environmental systems will need to be adequately designed to prevent the spread of cooking odours into other areas of the building.

Plant room

The plant room should be well maintained, clean and suitable for the size of the gym.

Storage

Adequate secure equipment storage is essential in order to provide a range of classes. For multipurpose studios, equipment may need to be stored or retrieved between classes quickly and with ease.

Staff facilities

These back of house areas will need to be appropriate to size of the facility.

Other facilities

In addition, the gym may provide additional facilities, including none, some or all of the following:

- retail outlets or concessions
- swimming, training or leisure pools
- health spas, e.g. saunas, steam rooms and pools
- health and beauty treatments, e.g. massage, relaxation, alternative therapies, hairdressing and manicure
- crèche
- squash courts
- tennis courts
- physiotherapy/sports injury clinics
- first-aid room.

Basic valuation methodology

As with all trading properties, the nature of each business is so varied that direct comparable evidence is difficult to find. A twenty-four-hour gym is unlikely to have the same business profile as a women-only gym or a fitness centre, even if the classes and facilities are very similar. In addition, the lack of gyms generally, and the lack of gym sales on the open market, tends to lead to a lack of comparable evidence that valuers can rely upon.

As such, the preferred method of valuation tends to be the 'profits method', although it must be stressed that comparable evidence is still important for each part of the valuation under the profits method. Comparable transactions where available will help define yield selection, while comparable trading profiles will help determine potential trading profitability for the property.

Example: Border's Gym

In this example, the gym has a rising number of members, from 1,850 in the first year to 1,944 in the third year, with membership fees increasing from £40.00 to £45.00 over the same period.

The typical monthly rate also increases from £42.50 to £45.00, while the attrition rate is stable at 35 per cent, indicating the gym is getting close to stabilization.

As is typical, the majority of income (approximately 74 per cent) comes from monthly membership fees, with personal training generating approximately 16 per cent of turnover.

The net operating profit for the property is ranging between 30.5 per cent and 33.1 per cent.

The property is valued using a 12.25 per cent capitalisation rate, reporting a gross value of £3.71 m.

Operational challenges – getting to the bottom line

When carrying out the valuation of a gym, getting to the sustainable EBITDA of the property is the immediate goal. It is usually performed through a number of quite separate exercises:

- property inspection
- analysis of the historic trading accounts
- review of the competition
- interview with the manager
- independent analysis of everything that has been seen and said throughout the process to determine how the gym is likely to trade in the hands of a reasonably efficient operator (REO).

Assessing the likely trading profile of the gym will require a complete understanding of the trading environment in which the facility sits. It is difficult

Table 15.4 Border's Gym trading projections

	2016		*2017*		*2018*	
Maximum membership capacity	2,500		2,500		2,500	
Average monthly fee	42.50		44.00		45.00	
Number of members	1,850		1,896		1,944	
Average joining fee ($)	40		43		45	
Attrition rate	35%		35%		35%	
Revenues ($)						
Membership fees	943,500.0	77.6%	1,001,220.0	74.0%	1,049,574.0	74.1%
Joining fees	25,900.0	2.1%	28,207.0	2.1%	30,613.0	2.2%
Personal training	146,000.0	12.0%	219,000.0	16.2%	227,760.0	16.1%
F&B	45,212.0	3.7%	46,568.4	3.4%	47,965.4	3.4%
Retail	32,542.0	2.7%	33,681.0	2.5%	35,028.2	2.5%
Day-visitor income (including classes)	12,457.0	1.0%	12,955.3	1.0%	13,343.9	0.9%
Other revenue	10,257.0	0.8%	11,077.6	0.8%	11,631.4	0.8%
Total revenue	1,215,868.0	100.0%	1,352,708.9	100.0%	1,415,916.0	100.0%
Total operational costs	759,917.5	62.5%	818,388.9	60.5%	856,629.2	60.5%
Gross operating profit	455,950.5	37.5%	534,320.0	39.5%	559,286.8	39.5%
Fixed costs ($)						
Property taxes	72,952.1	6.0%	75,140.6	5.6%	77,394.9	5.5%
Insurance	12,158.7	1.0%	12,523.4	0.9%	12,899.1	0.9%
Total fixed costs	85,110.8	7.0%	87,664.1	6.5%	90,294.0	6.4%
EBITDA ($)	**370,839.7**	**30.5%**	**446,655.9**	**33.0%**	**468,992.8**	**33.1%**

Table 15.5 Border's Gym income capitalisation method (in $s)

Stabilised EBITDA in present values	468,993	
Capitalised at 12.25%	8.163	3,828,513
Additional assets		0
Less income shortfall	120,490	120,490
Gross value		3,708,023
Less transaction costs	0.00%	0
Net value		3,708,023
Net value – SAY		3,710,000

to provide a comprehensive check list of things that need to be considered, as each gym tends to be quite different. However, breaking down the trading profile into sections is usually desirable, looking at the various income streams, followed by the various operational costs.

This is one of the most sensible approaches as it will usually result in a thorough analysis of the likely vagaries of that specific business.

Income

Typically, a gym will generate income from a number of different sources, and it is essential that the valuer reviews each of these in detail to be able to accurately project the likely income stream for the gym going forward.

These may include the following.

Membership

The majority of the income for most gyms comes from its membership income. Gyms may have a joining fee as well as a monthly membership fee. Most gyms will offer different membership packages including peak membership, off-peak packages, family membership, corporate membership and couple membership.

The better the quality, and more highly desirable the gym, the higher the potential number of members and the annual fee that can be charged.

Generally, the more facilities offered the higher the annual membership, and there are very distinct differences in membership costs for budget gyms, weight training gyms, gyms with wet facilities, and gyms with racquets/tennis facilities.

Personal training

Usually one of the most profitable areas of income, personal training is offered by the gym staff on an hourly basis, sold either as a one-off session or more commonly as part of a structured programme.

The income received from this source will be determined by the quality and the personality of the personal trainers, as well as by the sales training and skill of the whole team.

Food and beverage

This can be quite a sizeable income stream for a gym, as it is not unusual for gym members to use the facility after a workout session.

The level of the potential income will depend upon the membership profile, as well as careful planning of the F&B offering to tie in with the member's desires and requirements.

Retail

Gym shops can generate reasonable income streams depending on the membership base.

Additional offerings

Gyms often offer additional training options that are excluded from the 'classes' contained within the membership package. This might include tennis lessons, aquatic programmes or 'celebrity' classes.

Day-visitor fees

A number of gyms will allow day membership, charging accordingly for use on a daily basis.

Other income

There is usually a separate 'other income' line in the accounts to cover all income that is not covered by specific individual lines. It might include items such as rental income, advertising, crèche facilities or private hire, for example.

Costs

Costs are usually broken down into many areas including the direct cost of sales, staffing costs, sales and marketing, laundry, administration and general, repair and maintenance, water charges, insurance and taxes.

It is important to review the cost base so it reflects the likely profitability of the specific property. It may not be possible, for example, to meet typical F&B operating margins if the use is low, or the area is poorly designed and fitted out.

What are customers looking for in a gym?

All gym customers will have basic expectations of a gym, including cleanliness, friendliness and approachability of staff. However, there are a number of key factors that will differentiate customer satisfaction, and therefore the amount of repeat business generated at the property.

The customer comes to the gym for a number of key reasons, and a good gym will ensure their customer's requirements are met. As discussed earlier, the target audience of the gym will determine these key requirements, and although these items will vary for each type of gym, they will typically include the following.

Attaining individual goals

The single most important factor in customer satisfaction is achieving their personal targets. If a customer is looking to lose weight and they achieve this goal, they will recommend the gym to their contacts. If the goal is to enhance body mass and this is achieved, they are likely to remain loyal customers.

As such, having the most suitable equipment and the best staff (in terms of knowledgeable trainers, excellent motivators and having the ability to help set attainable goals) is the key to long-term success.

Pleasant environment

There are many factors that will determine whether the overall environment is appealing to customers. These will include:

- When people first walk in, it must be clear what kind of gym it is.
- The gym needs to be inviting to the clientele it is catering to. Cleanliness is a key indicator of a quality gym. It shows how much the staff and the owners care about their members.
- The locker room can be a big problem and a major turn-off to members if not kept clean.
- Quality gyms take steps to control odour. It is a gym and people sweat, but the gym itself should not stink.
- The gym should be well lit and should not be a gloomy environment.
- Air conditioning, air movement and low humidity are important for a quality gym. The standard temperature is between 20–22°C. It may be cold for some people but when you're working out, it won't feel cold for very long. Some gyms try to save money by raising the temperature but ultimately they will lose members!
- The number one reason people join a gym is to use the equipment and resources for the purpose of getting in shape. If the equipment is well maintained and clean, it makes for a good experience.
- The quality, variety and number of stations will impact on membership numbers. In addition, the layout and the space between machines will also impact on revenue streams.
- The quality of the staff is also important. Friendliness, levels of knowledge and their sociability all impact on customer satisfaction. This is as true for reception as it is for the personal trainers.

Value for money

Most customers feel the need to receive value for money, even at the very top end of the market where price sensitivity is less of an issue. The relative cost of the membership must reflect the quality being offered.

Inspection

A typical gym inspection will follow the general course of a hotel inspection. Initially, the valuer will request the information required, ahead of the inspection. This will include all, or some, of the following.

Legal information

- title documentation or reports on title
- ground leases or property leases
- subleases
- occupational licences or concessions
- planning consents for the site and each building on the property
- any restrictive covenants and other use prohibitions
- details of any unused planning consents
- details of any unfulfilled planning conditions
- copies of the operating licences
- agreements in place with personal trainers, masseurs or other professionals.

Property information

- area of the overall site
- site plans (with sizes) for each building
- building plans – floor by floor
- breakdown of gym facilities (by room or area)
- breakdown of other facilities, including F&B outlets
- copies of any current or historic valuation reports
- copies of any current or historic condition surveys, asbestos surveys, etc.
- details of historic capital expenditure on the properties (last three years)
- details of any proposed capital expenditure (this year and next year's budget).

Operational information

- historic profit and loss accounts (three years minimum)
- example of typical terms and conditions of the membership agreements
- management statistics, including membership numbers and rates as well as attrition rates
- statutory accounts
- full profit and loss budget for the current year and for the next year
- details of market segmentation and age/sex analysis of the members' profile

- membership numbers and rates, broken down by membership category
- staff details – numbers of staff and contracts of employment, including pension details
- length of service, and skills training for each member of staff
- sales and marketing plans; this year and next year's budget
- details of f and b revenues; where it comes from, average cover spends, etc.
- details of the local markets; SWOT analysis of the local competitors
- details of local trading; how does the property compete with its competitive set?
- details of operational equipment.

Other

- details of any outstanding complaints, legal claims etc.
- details of the tax positioning of the property (outstanding tax bills, capital allowances, etc.)

The gym will be inspected, usually in the presence of the manager (or department head if it forms part of a hotel or resort). The valuer will be looking at the gym facilities to see that it is suitable for use and how appealing it will be to the local market.

This will relate to the layout of the facilities, the classes on offer, the price of membership, as well as the reputation and quality of the personal trainers.

Specific care will be taken where the gym forms part of a resort or hotel to ensure it works effectively as a stand-alone unit, while seamlessly integrating with the larger property around it.

In addition, the valuer will be noting details about the following:

- Car parking – is it suitable for the volume of customers? Could it handle additional business without creating operational difficulties? Is there spare capacity that could be used for an alternative use?
- Non-income-generating space – are the relaxation areas, reception areas, swimming pools, saunas and steam rooms appropriate for the market needs?
- Rights of way – are there any public rights of way? Do they impinge on the operation of the business?
- Other facilities – are there any other areas that could generate income for the business? Are there any items that might raise costs for the business?
- Equipment – is all the equipment up to date and suitable for use? Does it have operational issues that need to be reflected in the trading profile? Will it need to be replaced soon?
- Health and safety – does the property comply with current and future legislation? If not, what are the cost implications?

In effect, the purpose of the inspection is to see all matters that may impact on revenue generation or have cost implications, so an accurate assessment of future trading can be made.

The condition of everything will also be assessed to ensure that any future capital expenditure requirements can be accurately gauged.

The quality of the gym is then reviewed in relation to the relevant market, so an accurate assessment of the appropriate yield for the property can be made to finally determine value.

Interview with the general manager

After the inspection of the property, the trading history, past, present and future will be discussed with the management.

Typically, a detailed discussion will review the operational issues relating to the gym in terms of revenue generation, followed by anticipated costs resulting in an understanding of the future trading profile and a review of the proposed capital expenditure plans.

The local market can be key to revenue generation for the property, with new supply potentially impacting on membership numbers. As such, a detailed discussion with the general manager about the various strengths and weaknesses of the gym, in relation to the existing (and proposed) market, is desirable to allow the valuer to understand the local market dynamics as seen from the perspective of the current manager.

The income generated by each segment, including membership numbers and rates, day use, F&B income and other revenue, will be discussed in light of past trading so an understanding of the projections can be fully made.

The same process will occur for each revenue line, followed by each cost line to ensure the trading projections are typical of the local market, and achievable.

The valuer will ask to review any licences that are required for operation, discuss the ownership of the property, ownership of the equipment and details of any franchises or concessions.

Example: Hadlee's Health Club

In this example, a large gym has a rising number of members, from 8,400 in the first year to 8,825 in the third year, with a maximum capacity of 10,000 members. Membership fees increase from $120.00 to £160.00 over the same period. The typical monthly rate also increases from $62.25 to $70.23, while the attrition rate is dropping, from 31 per cent down to 28 per cent.

As is typical, the majority of income (approximately 73 per cent) comes from monthly membership fees, with personal training generating approximately 11 per cent of turnover and F&B generating almost 6 per cent.

The net operating profit for the property is ranging between 34.1 per cent and 34.9 per cent.

The property is valued using a 15 per cent capitalisation rate, reporting a gross value of $22.95 m.

Table 15.6 Hadlee's Health Club trading projections

	2016		*2017*		*2018*	
Maximum membership capacity	10,000		10,000		10,000	
Average monthly fee	65.25		68.51		70.23	
Number of members	8,400		8,610		8,825	
Average joining fee (s)	120		150		160	
Attrition rate	31%		29%		28%	
Revenues ($)						
Membership fees	6,577,200	74.4%	7,078,712	72.9%	7,437,071	73.1%
Joining fees	313,488	3.5%	368,078	3.8%	395,371	3.9%
Personal training	821,250	9.3%	1,095,000	11.3%	1,138,800	11.2%
F&B	542,125	6.1%	558,389	5.8%	575,140	5.7%
Retail	195,125	2.2%	201,954	2.1%	210,033	2.1%
Day-visitor income (including classes)	310,254	3.5%	322,664	3.3%	332,344	3.3%
Other revenue	75,489	0.9%	81,528	0.8%	85,605	0.8%
Total revenue	8,834,931	100.0%	9,706,324	100.0%	10,174,364	100.0%
Total operational costs	5,318,628	60.0%	5,809,235	60.0%	6,089,357	59.9%
Gross operating profit	3,516,302.5	39.8%	3,897,089.2	40.2%	4,085,007.2	40.2%
Fixed costs ($)						
Property taxes	441,746.6	5.0%	454,998.9	5.0%	468,648.9	4.6%
Insurance	66,262.0	0.8%	68,249.8	0.7%	70,297.3	0.7%
Total fixed costs	508,008.5	5.8%	523,248.8	5.4%	538,946.3	5.3%
EBITDA ($)	**3,008,294.0**	**34.1%**	**3,373,840.5**	**34.8%**	**3,546,060.9**	**34.9%**

Table 15.7 Hadlee's Gym income capitalisation method (in $s)

Stabilised EBITDA in present values	3,546,061	
Capitalised at 15.00%	6.667	23,640,406
Additional assets		0
Less income shortfall		709,987
Gross value		22,930,419
Less transaction costs	0.00%	0
Net value		22,930,419
Net value – SAY		22,950,000

16 Golf course values and how they are determined

Introduction

This chapter will explain what makes a good golf course, how golf courses are valued, the sort of purchasers that exist for stand-alone golf courses and how golf courses can work effectively as part of a resort or attached to a hotel.

To paraphrase an old saying 'I can't define what makes a golf course great, but I know it when I see it'. This shows just how difficult it can be to assess what actually defines the quality of the golf course. Is it the setting (Pebble Beach or Gleneagles)? Is it the beauty (Augusta or San Lorenzo)? Is it the challenge (St Andrews Old Course, Pine Valley or Les Bordes)?

There has been a recent trend in golf design to simple equate the length of the holes with the quality of the course, which is too simplistic. Most golf courses will have their own personalities, which is essential to generating loyalty for members and visitors alike, over and above the pre-requisites of good functional design.

An eighteen-hole course will typically be a series of three, four and five-par holes, leading a total par score over the eighteen holes of around seventy-two. Each hole will have a series of tee points (different for professionals and amateurs and for women and men), a central fairway, with an area of rough on both sides, followed by a putting green where the hole is located. There may be any number of hazards on the course including woodland, out of bounds areas, sand traps or water features. The layout of each hole, the size and shape of the fairway, the approach to the putting green, the surrounding landscape, the topography and the 'plants' all help set the context for the course's 'personality'. The more personality a course has, the more desirable it can be for both customers and potential owners.

This 'uniqueness' attracts certain types of investor into the golf world, making it quite different from standard commercial property. There is an old adage in golf course development that new courses are developed by passionate individuals who go bankrupt before completion. The half-built course is started by an enthusiast who cannot make a return on the investment and is forced to sell. It is then purchased by a mainstream operator at a realistic price and they are finally able to make a realistic and sustainable profit from the business.

This might be very simplistic, but historically there is some truth behind the adage. It is therefore helpful to look at the typical purchasers of golf courses to learn about the motivations, which in turn are instructive when assessing market value.

Golf courses can work extremely well as part of a larger operational entity, for example, a hotel or a resort. Demand for other services can be generated by golfers, and as well as operational costs being shared with other parts of the operation, reducing the impact of such costs, to the benefit of the overall profit margins. However, in some instances, generally where the golf is not generating enough demand or it is badly run, the golf element can be a drain on an otherwise successful hotel or resort business.

Purchasers for resorts featuring golf courses and those for golf hotels tend to be different to purchasers of stand-alone golf courses, so although the methodology for assessing the operating profit will be similar, the yield profile may be different. As such, it is quite important to assess likely purchasers as an integral part of the valuation process.

Types of buyers and motivations for ownership

The market for golf courses is not as widespread as it is for hotels, and when the market for a particular property is very narrow, understanding the detailed motivations behind a potential purchaser's acquisition program is essential to assessing value.

The main categories of purchasers for golf courses can be summarised as follows.

Major commercial operators

The typical motivation for major commercial operators is to generate financial profits from the business. They will typically have expectations of financial returns, and unless there is substantial room for trading enhancement or they have a business requirement for that location, they tend to be careful to ensure the yield profile of any purchase suits their particular business model.

Typical golf courses that would appeal would have strong membership lists, a track record of good trading (compared with the price being paid) and would usually have a minimum of eighteen to twenty-seven holes, a good-sized clubhouse, driving range, putting green and a retail outlet.

Small commercial operators

Small commercial operators are also typically attracted to the potential financial returns available in operation of the course, though they tend to be more geographically limited and attracted by opportunities close to their existing course or courses.

Pricing may be dependent on their ability to raise bank debt to finance the purchase and can be determined by the potential synergies between the new course and existing businesses, and any potential cost savings.

The typical minimum requirement would be eighteen holes and a clubhouse for such purchasers, though properties with only nine holes can sometimes appeal as will larger properties, depending on the exact nature of the potential purchaser.

Lifestyle buyers

As is the case with hotels, lifestyle buyers tend to be attracted to golf courses for the quality of the environment and the potential lifestyle from running a golf course. The purchasers tend to be enthusiastic golfers and they will sometimes pay prices that represent the level of their desire for the property, rather than what makes financial sense in terms of likely returns.

In much the same way a house buyer who falls in love with the view from a house is prepared to pay a price that others in the market may not be prepared to pay, a lifestyle buyer can sometimes be enticed to pay more money than others in the market would deem sensible.

Trophy buyers

As is the case with hotels, trophy buyers tend to purchase trophy courses because they wish to own them, rather than because of any operational or financial return requirement. Prices paid tend to include a premium, if it has been marketed to the 'right' buyers and in the 'right' way. The level of that 'premium' will depend on the course, as well as the market conditions at the time of sale.

A trophy course may be a championship course, a historic course with grand associations, one of the top courses in a given location or it may be a particularly well regarded or photogenic course. There is likely to be an extensive membership list, and potentially a waiting list for membership, as well as full services (putting green, driving range, clubhouse, shop, coaching, etc.) provided within in the grounds.

Entrepreneurs and developers

Certain clubs appeal to entrepreneurs when they are considered 'undervalued'. This might be the case where a local business person believes the operation or marketing of the course has not been optimised, and therefore there is potential 'upside' for the business trading that is not being factored into the pricing of the property.

Developers also fit into this category, and typically they believe that there is latent value in the site that they can exploit. This may potentially be generated by developing residential properties on the site, or by selling off parcels of land or properties that are not required for the operation of the course.

Private members clubs

There is also one other key group of potential purchasers: the general playing public or members of the existing or competing courses. It is not uncommon for such groups to form into a type of 'ownership entity' with the intention of buying a club (especially if it is in financial difficulty) to protect the club from liquidation, or to change the 'direction' of the management of the club.

There are many types of entity formed, including partnerships, limited companies and even trusts. The final entity is relatively immaterial; the key factor is the motivation behind their purchase.

Characters of golf

Chris Honeywill, director – Lambert Smith Hampton

At first I was going to discuss the benefits that a golf course can bring to a resort. This can be put simply – you build a golf course in the right place, which creates the destination; build a new hotel to accommodate visiting golfers; create a commercial area with shops, restaurants, bars, etc.; and then sell plots of land, apartments or villas. It only needs the right site; planning consent; entrepreneurial and development skills; solid marketing and a good economy. It is quite straightforward, so I have wandered elsewhere.

My decades in the property industry have largely been spent in the UK, where specialist valuers will tell you that golf properties are assessed on a trading-related basis. Typically, EBITDA will be 20 per cent of turnover and the capital value of the freehold is based on eight to ten-times EBITDA, give or take – plus any clear development potential.

However, in my experience the 'characters' of the golf business have a better feel than the technical approach. The secret seems to be having the ability to spot the potential and know where the income is coming from. Albeit that the banks lend money based upon those technical appraisals.

The late Ron Noades saw the advantage of creating income when it was dark or wet. He invested heavily in clubhouse facilities, compared to others, and the function/hospitality side is still very successful, as well as being pretty good for golfers. The car parks are rarely empty.

Others have created strong cash receipts from different opportunities, such as landfill (whoops – using inert waste for course improvements) or developing the odd house or two (manor from heaven). Trophy sites like Wentworth have changed hands at huge premiums – £135 m on an EBITDA reported to be £7.5 m and the investors are labelled as 'characters'.

The following is a true story about two business characters.

My first golf instruction came along in 1981. Silvermere Golf and Country Club had fallen into difficulty and my firm was instructed to sell by the receiver. I was sent along, being the only golfer in the firm. Maybe a daft reason, but I was very happy.

Silvermere was a strange set-up. Eighteen difficult golf holes built on poor-draining land, which opened in 1976 – one of the hottest summers on record. During wet spells, the lower holes became very 'boggy' and occasionally golfers actually lost their shoes! Not surprisingly, the golf side struggled. Also, there was an equestrian centre with a three-day eventing course around the site boundary, taking up twenty acres, plus a dressage square. There was a swimming pool, tennis courts and a collection of cheap buildings housing the restaurant/bar, changing rooms and a ladies' hairdresser.

We quickly sold the property to a contact of our late senior partner Robert Sice, and Ronnie Shaw bought the property for £850,000. Shawline

Investments had a range of property interests, mainly in the medical sector in London but also 50 per cent of Mijas Golf in Spain. Ronnie's main love was for racehorses, so it was a surprise that the first thing he did was to sell the equestrian centre for housing development. The membership was closed and the tennis and swimming pool were removed. Doug McClelland, Master PGA professional, was put in charge. The hairdresser left.

In 1984, Ronnie wanted cash to help him set up a horse racing track in Marbella and we were asked to sell Silvermere quickly. I started to prepare a marketing campaign to sell by auction but knew that a good friend, Tom Hilliard, wanted to diversify from his fruit wholesale business. After a speedy introduction, Ronnie asked Tom why he wanted to buy a golf course and Tom answered, 'to lower my golf handicap'. They shook hands at £1.025 million.

Tom and Doug spent the following twenty years developing Silvermere as one of the best businesses in the golf industry in Europe. Huge investment into the course overcame the drainage problems. The golf store has an enviable layout and is the busiest of golf retail outlets; recently the driving range has been rebuilt as a fifty-six-bay, two-tier unit and that followed on from the new clubhouse with an F&B side to meet all tastes. Turnover has increased many times the price Tom paid due to their energy, work ethic and confidence to invest. Being located at the junction of the M25 and A3, close to Brooklands Business Park and many 'chimneypots' may have helped. Perhaps Tom spotted the potential the location held.

Neil Coles once said that a successful golf facility is usually created after it has gone bust first time round. There are many examples that prove that maxim, including Silvermere. Sadly, there are examples of golf courses reverting back to agricultural use or even the odd one getting planning consent for housing. Recently, the renowned Brocket Hall and Mentmore clubs have fallen into administration but perhaps someone like Tom will arrive before the grass grows too long.

Currently, local suburban clubs find it hard. Joining fees having dropped away, waiting lists are a thing of the past and new members are readily welcomed. Against this, the top-end market thrives, such as the exclusive Queenwood being sold out at $250,000 entry cost and the new Beaverbrook Club in Leatherhead having sold some 220 memberships where the entry fee now stands at £145,000, which is not bad as it will not open until 2016. Perhaps the men behind those schemes will become the future characters of the industry.

It is a long time since the 1989 call from the R&A for 500 more golf courses – it is a pity they didn't add 'for "pay and play" purposes or to suit millionaires, but nothing in between'. The uncertainty behind the trading potential of many facilities is perhaps the reason why I have dealt with more than 350 golf instructions since 1981 – and long may that continue.

The likely profile of potential purchasers will therefore sometimes have a major impact on the methodology applied to the valuation, as market value can reflect the way buying decisions are made by each group.

When a golf course forms part of a resort or a hotel, it tends to attract typical hotel or resort purchasers, and the motivation for purchase (whether for financial returns, a trophy purchase, to have another 'flag' or for financial security) is unlikely to be radically different because it includes a golf course.

Basic valuation methodology

As with all trading properties, the nature of each business is so varied that direct comparable evidence is not easy. A business that has a direct catchment area of 400,000 is unlikely to have the same business profile as one with a 50,000 catchment area, even if the overall membership size is similar. In addition, the lack of golf courses generally, and the lack of golf course sales on the open market, tends to lead to a lack of comparable evidence that valuers can rely upon.

As such, the preferred method of valuation tends to be the 'profits method', although it must be stressed that comparable evidence is still important for each part of the valuation under the profits method. Comparable transactions where available will help define yield selection, whilst comparable trading profiles will help determine potential trading profitability for the property.

Example: Royal Lillee Links

In this example, the golf course is projected to grow membership from 2,500 to 2,580, playing between 22,500 and 22,639 rounds per year, with average green fees ranging from $65 to just under $71.

Profitability is forecast to increase slightly over the first three years, as cost savings are anticipated.

In addition, the property has surplus property that can be sold off without impacting the operation of the course which is worth an additional $450,000.

Operational challenges – getting to the bottom line

When carrying out the valuation of a golf course, getting to the sustainable EBITDA of the property is the immediate goal. It is usually performed through the following separate exercises:

- property inspection
- analysis of the historic trading accounts
- review of the competition
- interview with the general manager
- independent analysis of everything that has been seen and said throughout the process to determine how the golf business is likely to trade in the hands of a reasonable efficient operator (REO).

Table 16.1 Royal Lillee Links projections in future values

	2016		2017		2018	
Members	2,500		2,550		2,580	
Membership fee ($)	130		140		145	
No. of rounds	22,500		23,063		23,639	
Average round fee ($)	65.00		68.25		70.64	
Revenues ($ '000s)						
Membership fees	325	14.6%	357	15.0%	374	14.9%
Green fees	1,463	65.8%	1,574	66.0%	1,670	66.4%
Range	65	2.9%	67.4	2.8%	70.1	2.8%
Retail	85.8	3.9%	88.3	3.7%	91.0	3.6%
Miscellaneous golf	23.1	1.0%	23.9	1.0%	24.9	1.0%
Total F&B	175.5	7.9%	182.5	7.7%	187.9	7.5%
Other revenue	85.1	3.8%	91.9	3.9%	96.5	3.8%
Total revenue	2,222.4	100.0%	2,385.1	100.0%	2,514.4	100.0%
Operational costs ($ '000s)						
Direct cost of sales	107.9	4.9%	113.3	4.9%	117.9	4.9%
Golf wages	437.6	19.7%	450.7	19.7%	464.3	19.7%
F&B wages	61.4	2.8%	63.9	2.7%	65.8	2.7%
Admin wages	277.7	12.5%	286.0	12.5%	294.6	12.5%
Golf other	265.8	12.0%	273.8	12.0%	282.0	12.0%
F&B other	30.4	1.4%	31.3	1.4%	32.3	1.4%
Admin and sales other	460.7	20.7%	474.5	20.7%	488.7	20.7%
Other costs	15.5	0.7%	15.9	0.7%	16.4	0.7%
Total operational costs	1,657.0	74.6%	1,709.5	71.7%%	1,761.9	70.1%
Gross operating profit	565.4	25.4%	675.6	28.3%	752.5	29.9%
Fixed costs ($ '000s)						
Property taxes	85.0	3.8%	87.6	3.7%	90.2	3.6%
Insurance	22.2	1.0%	22.9	1.0%	23.6	0.9%
Ground rent	12.0	0.5%	12.4	0.5%	12.7	0.5%
Total fixed costs	119.2	5.4%	122.8	5.1%	126.5	5.0%
EBITDA ($ '000s)	**446.2**	**20.1%**	**552.8**	**23.2%**	**626.0**	**24.9%**

Assessing the likely trading profile of the golf course will require a complete understanding of the trading environment in which the course sits. It is difficult to provide a comprehensive check list of things that need to be considered, as each

Table 16.2 Royal Lillee Links income capitalisation method (in $s)

Stabilised EBITDA in present values	625,983	
Capitalised at 15.00%	6.667	4,173,221
Additional assets		450,000
Less income shortfall		
Gross value		4,370,211
Less transaction costs	6.00%	262,213
Net value		4,107,998
Net value – SAY		4,100,000

course tends to be quite different. However, by breaking down the trading profile into sections, it is usually easier. Looking at the various income streams, followed by the various operational costs, is one of the most sensible approaches as it will usually result in a thorough analysis of the likely vagaries of that specific business.

Income

Typically, a golf course will generate income from a number of different sources, and it is essential that the valuer reviews each of these in detail to be able to accurately project the likely income stream for the course going forward. These may include the following.

Membership fees

These are sometimes also called 'green cards' or 'membership subs' or just 'subscriptions'. This is the annual fee paid by members to join the club. It will also include life membership fees, though of course these tend to be a one-off fee, so are not typically replicated every year.

Membership fees will be determined by the number of members and the fees they are prepared to pay. The more desirable or prestigious the club ,the higher the fee levels that can be charged. Some clubs will be operating at capacity and cannot have any additional members so the only way for growth would be in price increases.

It is sometimes possible to enhance membership numbers by offering different categories of membership, to enhance play through different times of the day and year, without impacting on demand at peak periods.

Green fees

Green fees are the charge for use of the golf course. Charges are levied upon use of the course, and green fees are determined by the amount of rounds played on the course each year, as well as on the cost of each round.

Green fees may vary by day, by time of play or by season, with some clubs enhancing revenue through careful matching of prices to the strength of demand.

Driving range

Most golf courses provide a driving range for members to use. Typically, charges are incurred based on the number of balls that are used or on the length of time that a driving bay is occupied. This potential income stream can be enhanced by the provision of floodlighting, allowing the range to be used for longer periods of time.

Shops

Golf shops can generate quite significant income streams, especially in courses with significant non-member use. In certain clubs, the space is leased out to third parties which takes away the operational risk and provides a rental income to the club instead. However, the rental income is usually lower than the profit that would typically be generated by a well-run retail unit.

Golf miscellaneous

This is additional golf-related income that is not covered in the various separate income lines elsewhere in the accounts. It will usually include any number of things, but typically will include buggy hire, golf club hire and commissions from any coaching arranged with the golf professional.

Clubhouse (or F&B)

This may or may not be broken down into food and beverage departments, and it may even be broken down by outlet if the club has more than one facility. Usually any meeting room income, private hire fees and any wedding income is included under this department.

A typically well-located clubhouse will have views out onto the eighteenth green, so friends, families and other players can be spending money at the bar whilst watching the end of a round.

The size of the clubhouse, its facilities, design and decoration, the number of members, the number of competitions held at the course, along with the friendliness of the staff and speed of service will all have a direct impact on the revenue generation of the department.

It may be that the clubhouse facility has been let out to a third party or is operated under licence. This will have a direct impact on the income and cost implications for that department.

Other income

There is usually a separate 'other income' line in the accounts to cover all income that is not covered by specific individual lines. It might include items such as rental income, advertising, or private hire, for example.

Costs

Costs can be broken down into many areas including the direct cost of sales, staffing costs, sales and marketing, administration and general, repair and maintenance, water charges, insurance and taxes.

Special care should be taken to ensure all operating costs are accurately accounted for, including rectifying or preventing industry specific problems like fly tipping or flooding.

It is important to review the cost base so it reflects the likely profitability of the specific property. It may not be possible, for example, to meet typical F&B operating margins if the kitchen is poorly designed and fitted out. The cost of sales for the driving range could potentially be higher than industry averages if retrieving the balls is more complicated than usual.

There is one key area where careful consideration has to be made. If a golf course is used for competition play, it can have an impact on the amount of public use it can have. Championship courses usually generate lower green usage than non-championship courses, though of course championship courses can sometimes attract greater green fees and will sometimes attract sharper yields if the property is sold.

What makes a golf course popular?

It is quite interesting to be aware that there is little definitive data on what golfers actually want from a golf course – the key to finding out what makes a customer happy. If a club can make its customers happy then the business should be successful, as long as there is a sensible business model in place.

One key piece of research was undertaken by Sachau, Simmering and Adler in *Golf Digest* 2013 to determine what was important for golfers in their experience of visiting a new venue (Figure 16.1). The study was based on 'away play' as there was a definitive belief that the overriding single most important factor determining satisfaction levels at the 'home venue' was the speed of play. If a player was held up by slow players ahead, or felt rushed by others behind, it provided the lowest satisfaction rating of any single factor. The logic for this was that 'at home' you had limited time to play, after which you needed to carry on with the usual tasks of your daily life. The study posited that an 'away' visit had potentially different and more interesting priorities.

The study determined that despite what the players said, pace of play was not actually a very important determinant on whether they were satisfied with the golf experience at an away course. Overall, it found that what golfers said was most important was not always accurate when looking at what actually made them satisfied.

The important thing for the valuer to take from this, and indeed anyone who is hoping to run a successful golf operation, is that the overriding concern of most players is the quality of the greens and the green fees. Although green keepers are back of house staff, they have the highest impact on player satisfaction, and therefore this is one of the best places for an investment return.

Figure 16.1 What golfers *say* makes them happy and what *actually* makes them happy

Inspection

A typical golf course inspection will follow the general course of a hotel inspection. Initially, the valuer will request the information required, ahead of the inspection. This will include all, or some, of the following:

Legal information

- title documentation or reports on title
- ground leases or property leases
- subleases
- occupational licences or concessions
- planning consents for the site and each building on the property
- any restrictive covenants and other use prohibitions
- details of any unused planning consents
- details of any unfulfilled planning conditions
- copies of the operating licences
- agreements in place with golf professionals, etc.

Figure 16.2 Golf hotel

Figure 16.3 Golf course

Property information

- area of the overall site
- site plans (with sizes) for each building
- building plans – floor by floor
- breakdown of golf facilities (by hole)
- breakdown of other facilities including F&B outlets, etc., by number and size
- copies of any current or historic topographical surveys
- copies of any current or historic valuation reports
- copies of any current or historic condition surveys, asbestos surveys, etc.
- details of historic capital expenditure on the properties (last three years)
- details of any proposed capital expenditure (this year and next year's budget)
- details of ground-keeping equipment, separating leased and owned property
- details of unused operational property, for example residential properties or equipment sheds.

Operational information

- historic profit and loss accounts (three years minimum)
- statutory accounts
- full profit and loss budget for the current year and for the next three years (if possible)
- details of market segmentation and geographical analysis of the members and guest profile
- details of lifetime members
- membership numbers and rates, broken down by type of membership
- staff details – numbers of staff and contracts of employment, including pension details, length of service and skills training for each member of staff
- sales and marketing plans; this year and next year's budget
- details of F&B revenues; including average cover spends, etc.
- details of the local markets; SWOT analysis of the local competitors
- details of local trading; how does the property compete with its competitive set?
- details of operational equipment.

Other

- details of any outstanding complaints, legal claims etc.
- details of the tax positioning of the property (outstanding tax bills, capital allowances, etc.)

The course will be inspected, usually in the presence of the general manager or the head green keeper. The valuer will be looking at the course to see that it is suitable for use and how appealing it will be to the local market. This may relate to the layout of the holes, who the designer was, the quality of the greens, the hazards and water features or even the length of the holes and its difficulty level.

Specific care will be taken where more than one course is present (whether a full course and a nine-hole course, or two eighteen-hole courses) to ensure they interact well without interfering with play on either course.

In addition, the valuer will be noting details about the following:

- Car parking – is it suitable for the volume of customers? Could it handle additional business without creating operational difficulties? Is there spare capacity that could be used for an alternative use?
- Irrigation – is it suitable for the course needs? How old is it? How efficient is it? Does it need upgrading? Does it cover the whole course?
- Drainage – how adequate is the drainage system and does the sub-soil cause a problem? Can anything be done to rectify any issues, or do any failings impact on the revenue generation potential of the course?
- Practice facilities – how do these compare with the competition? Are they adequate for the current market place? Could more members be attracted by providing additional facilities?
- Green keeping compound – is it suitable for the course? Where is the compound located? How large is it? How secure is it?
- Services – is the course on mains electricity, water, sewage and gas?
- Rights of way – are there any public rights of way? Do they impinge on the operation of the business?
- Clubhouse – where is it located? How large is it? How well designed is it? How well is it furnished? Does it meet with the customers' requirements? How does it compare to the competition? Can it be used for other events, such as weddings, Christmas parties, etc.?
- Other facilities – are there any other areas that could generate income for the business? Are there any items that might raise costs for the business?
- Equipment – is the green keeper's equipment suitable for use? Does it have operational issues that need to be reflected in the trading profile? Will it need to be replaced soon?
- Health and safety – does the property comply with current and future legislation? If not, what are the cost implications?

In effect, the purpose of the inspection is to examine all matters that may impact on revenue generation or have cost implications, so an accurate assessment of future trading can be made.

The condition of everything will also be assessed to ensure that any future capital expenditure requirements can be accurately gauged.

The quality of the course is then assessed in relation to the relevant market, to ascertain the appropriate yield for the property and to finally determine value.

One other key feature of the inspection will be to check for alternative uses that might have value implications. For example, the course might incorporate residential accommodation on site that is not needed for the operation of the course which could potentially be sold off without any adverse impact on the trading business.

Interview with the general manager

After the inspection of the property, the trading history, past, present and future, will be discussed with the management.

Typically, a detailed discussion will review the operational issues relating to the course in terms of revenue generation, followed by anticipated costs resulting in an understanding of the future trading profile and a review of the proposed capital expenditure plans.

The local market can be key to revenue generation for the property, with new supply potentially impacting on membership numbers and green fees. As such, a detailed discussion with the general manager about the various strengths and weaknesses of the course, in relation to the existing (and proposed) market is desirable to allow the valuer to understand the local market dynamics as seen from the perspective of the current manager.

Existing membership numbers compared with the historic numbers and future membership projections will be discussed, in light of existing and future market supply. Membership fees will be discussed in the same light, along with the structuring of membership packages. Details will be sought on 'life memberships' where no future income can be expected from members.

The number of rounds predicted to be played, along with overall projected green fees, will be discussed in view of past trading so an understanding of the projections can be fully made.

The same process will be used for each revenue line, followed by each cost line to ensure the trading projections are typical of the local market, and achievable.

The valuer will ask to review any licences that are required for operation, discuss the ownership of the property, ownership of the equipment and details of any franchises or concessions.

Example: Wasim Willows

In this example, the golf course is projected to grow membership from 125 to 130, playing between 10,450 and 10,979 rounds per year, with average green fees ranging from $85.00 to $92.37.

Profitability is forecast to increase slightly over the first three years, as cost savings are anticipated (Table 16.3).

However, in this example, the accounts have made allowance for a number of operational factors that are not usually deducted to calculate EBITDA: in this example £112,000 worth of additional expenses, including directors' remuneration, a business car used for directors of the club and an accounting expense that is not likely to be replicated by a potential purchaser.

As such, these expenses have been added back into the accounts to calculate an accurate EBITDA.

In addition, the property has surplus property that can be sold off without impacting the operation of the course which is worth an additional $100,000.

The ultimate golfing holiday

Claire Taylor, fanatical amateur golfer

As a 24-handicap lady golfer playing most weekends in club golf, I really enjoy my golf holidays. Having played golf at several holiday resorts, I have a good idea of what I like most. Bear in mind that holiday golfers can be a varied crowd, for example, club members, low-to-high handicappers, those who play regularly or those who play one or two times per year. Resorts that feature a variety of courses have the luxury of providing a selection of difficulties to suit all, which means golfers stay on site and spend their money, rather than exploring other locations nearby and taking their business elsewhere.

Most courses will be designed to avoid bottlenecks, provide a range of stunning views and toilet facilities as courses tend to be long, and most golfers take advantage of the beer buggy on the way round! It's important to stay hydrated, particularly in hot climates. Ultimately, I'm looking for a course which will challenge me but not be so difficult that I lose the holiday spirit. I want to play on well-manicured greens, encounter a range of hazards, such as water features, bunkers, humps and bumps, a mix of narrow and wide fairways to benefit the accurate and not so accurate, to encourage competition across the range of golfers in the typical group, and enjoy some 'wow' factors to share stories with my friends.

Excellent changing facilities, locker rooms, friendly staff, bar and snack areas and buggies are all key 'accessories' that make the day perfect, and a well-stocked pro-shop will most likely lead to purchases of branded items as a memento of the day or holiday.

Table 16.3 Wasim Willows projections in future values

	2016		*2017*		*2018*	
Members	125		128		130	
Membership fee ($)	130		140		145	
No. of rounds	10,450		10,711		10,979	
Average round fee ($)	85.00		89.25		92.37	
Revenues ($ '000s)						
Membership fees	16	1.5%	18	1.6%	19	1.5%
Green fees	888	82.6%	956	83.0%	1,014	83.4%
Range	45	4.2%	46.4	4.0%	48.2	4.0%
Retail	25.0	2.3%	25.8	2.2%	26.5	2.2%
Miscellaneous golf	24.0	2.2%	24.8	2.2%	25.8	2.1%
Total F&B	62.5	5.8%	65.0	5.6%	67.0	5.5%
Other revenue	14.3	1.3%	15.4	1.3%	16.2	1.3%
Total revenue	1,075.3	100.0%	1,151.2	100.0%	1,216.7	100.0%
Operational costs ($ '000s)						
Direct cost of sales	24.5	2.3%	25.7	2.3%	27.0	2.3%
Golf wages	232.0	21.6%	241.3	21.6%	253.3	21.6%
F&B wages	21.9	2.0%	22.8	2.0%	23.4	2.0%
Admin wages	125.4	11.7%	129.8	11.7%	134.3	11.7%
Golf other	110.3	10.3%	115.8	10.3%	121.6	10.3%
F&B other	9.2	0.9%	9.5	0.9%	9.8	0.9%
Admin and sales other	139.6	13.0%	146.6	13.0%	153.9	13.0%
Other costs	127.5	11.9%	131.3	11.9%	135.2	11.9%
Total operational costs	790.3	73.5%	822.7	71.5%	858.6	70.6%
Gross operating profit	284.9	26.5%	328.5	28.5%	358.1	29.4%
Fixed costs ($ '000s)						
Property taxes	85.0	7.9%	87.6	7.6%	90.2	7.4%
Insurance	10.8	1.0%	11.1	1.0%	11.4	0.9%
Ground rent	12.0	1.1%	12.4	1.1%	12.7	1.0%
Total fixed costs	107.8	10.0%	111.0	9.6%	114.3	9.4%
EBITDA ($ '000s)	**177.2**	**16.5%**	**217.6**	**18.9%**	**243.8**	**20.0%**

Table 16.4 Wasim Willows adjusted projections in future values

	2016		2017		2018	
Members	125		128		130	
Membership fee ($)	130		140		145	
No. of rounds	10,450		10,711		10,979	
Average round fee ($)	85.00		89.25		92.37	
Revenues ($ '000s)						
Membership fees	16	1.5%	18	1.6%	19	1.5%
Green fees	888	82.6%	956	83.0%	1,014	83.4%
Range	45	4.2%	46.4	4.0%	48.2	4.0%
Retail	25.0	2.3%	25.8	2.2%	26.5	2.2%
Miscellaneous golf	24.0	2.2%	24.8	2.2%	25.8	2.1%
Total F&B	62.5	5.8%	65.0	5.6%	67.0	5.5%
Other revenue	14.3	1.3%	15.4	1.3%	16.2	1.3%
Total revenue	1,075.3	100.0%	1,151.2	100.0%	1,216.7	100.0%
Operational costs ($ '000s)						
Direct cost of sales	24.5	2.3%	25.7	2.3%	27.0	2.3%
Golf wages	232.0	21.6%	241.3	21.6%	253.3	21.6%
F&B wages	21.9	2.0%	22.8	2.0%	23.4	2.0%
Admin wages	125.4	11.7%	129.8	11.7%	134.3	11.7%
Golf other	110.3	10.3%	115.8	10.3%	121.6	10.3%
F&B other	9.2	0.9%	9.5	0.9%	9.8	0.9%
Admin and sales other	139.6	13.0%	146.6	13.0%	153.9	13.0%
Other costs	15.5	1.4%	15.9	1.4%	16.4	1.4%
Total operational costs	678.3	63.1%	707.3	61.4%	739.8	60.8%
Gross operating profit	396.9	36.9%	443.9	38.6%	476.9	39.2%
Fixed costs ($ '000s)						
Property taxes	85.0	7.9%	87.6	7.6%	90.2	7.4%
Insurance	10.8	1.0%	11.1	1.0%	11.4	0.9%
Ground rent	12.0	1.1%	12.4	1.1%	12.7	1.0%
Total fixed costs	107.8	10.0%	111.0	9.6%	114.3	9.4%
EBITDA ($ '000s)	**289.2**	**26.9%**	**332.9**	**28.9%**	**362.6**	**29.8%**

Table 16.5 Wasim Willows income capitalisation method (in $s)

Stabilised EBITDA in present values	362,594	
Capitalised at 18.00%	5.556	2,014,411
Additional assets		100,000
Less income shortfall		103,100
Gross value		2,011,311
Less transaction costs	6.00%	120,679
Net value		1,890,632
Net value – SAY		1,900,000

17 Spa values and how they are determined

Roger Allen, CEO, Resources for Leisure Assets, and David Harper

Introduction

This chapter provides guidelines for valuation of a spa as an integral part of a hotel or resort development and as a stand-alone entity when using the profits method. Value determination of a spa can be a complex process, given the disparity in different types of spa facilities, categories, business models, organisational design and management performance. With the objective to simplify this complexity, background information on the most common types of spas along with profit and loss statement examples are provided from a hotel spa and as a stand-alone spa facility.

Over the last three decades, the global spa industry and the business of spas have grown dramatically. In the US alone, the number of spas has risen almost 110 per cent in the last decade.[1] Now it is almost inconceivable that a new hotel/resort of a four-star rating and above would be developed without including a spa or a spa-style element in the overall design. What started out as a differentiating factor for hotels and resorts in the early 2000s is now an international industry standard.

Considering the phenomenal growth of the spa industry worldwide, paradoxically, there is no simple, industry-wide definition of what a spa actually is. In several parts of the world, a spa can be merely a combination of beauty and body treatment options, in other parts of the world it requires water-based services, while in many countries, natural springs and thermal waters are an integral element, without which the property would not be considered a spa. According to the International Spa Association (ISPA), 'Spas are places devoted to overall well-being through a variety of professional services that encourage the renewal of mind, body and spirit'.

A spa within a hotel

The well-being expectations of today's travel and tourism market are driving a higher demand for hotel attractions and service provisions that meet a wide range of health, lifestyle and well-being needs of hotel guests. Spas can be highly beneficial for hotels and resorts, simply in meeting the needs of today's demanding traveller, as well as positioning for competitive differentiation in addition to revenue generation.

There is anecdotal evidence that states spas in luxury hotels can influence brand positioning, enhance upselling and trading across other operating departments, as well as generating a reasonable stand-alone operating profit when the spa is well planned, marketed and operated correctly. In addition, it is also suggested that spas provide options for travelling partners who might otherwise be reluctant to accompany their spouses on a business trip. Offering diverse activity and service venues for the double occupancy market can generate additional revenue streams for hotels and resorts, as well as potentially increasing the average length of stay.

Considering the fact that there are many types of spa business models, a hotel spa, when considered like any other operational department, should operate as a business to generate profits. The more profits the spa can contribute to the property, the more valuable it is as a business. However, different spas will vary in levels of investment required which will impact the earnings potential and overall real estate value of the property.

There is no single 'cookie cutter' hotel spa concept. Therefore, it is important to understand that different types of hotel properties and different markets will have differentiating spa facilities with operational variations. Concept, services and facilities of a spa can vary significantly depending on the owner's and/or operator's interpretation. For example, indoor/outdoor swimming and experience pools and exercise facilities may or may not be considered part of the spa.

The varying levels of investment, income potential and operational costs illustrate the challenges that need to be assessed. It should not be forgotten that the primary purpose of a hotel spa is to meet the needs of the hotel guest, sometimes over and above the need to generate departmental profits.

Two of the most common types of hotel property where a spa is likely to be found are as follows:

Hotel/resort spa

According to ISPA, a hotel/resort spa is categorised as a spa located within a hotel or resort that provides professionally administered spa services, fitness and wellness components.

The hotel/resort spa tends to encompass a dedicated area within the hotel or resort that is set aside for the exclusive use of spa facilities and services for hotel guests and in certain properties, membership options for local residents.

Typically, hotel/resort spas tend to be 'relaxation and pampering' spas, providing wet facilities such as experience pools, swimming pools, saunas, steam rooms and treatments such as personalised massages, body scrubs, facials and hydrotherapy treatments.

Spas located in city hotel properties and in those locations where sufficient residential populations exist have the foundations for offering membership schemes that may or may not also be part of a wider health club operation.

Destination spa

A destination spa is a facility with the primary purpose of guiding individual spa-goers to develop healthy lifestyle habits. This lifestyle transformation can be accomplished by providing a comprehensive programme that includes 'spa services, physical fitness activities, wellness education, healthful cuisine, and special interest programming', according to ISPA.

Guests typically stay at a destination spa for a number of days or weeks with the spa being an essential part of the facility. Destination spa facilities and services are typically more expansive in comparison to a hotel/resort spa with programmes offered for guests that may include detoxification, stress reduction, weight loss and behaviour modification, in addition to personal consultations with fitness, nutrition and medical professionals.

There are a large number of destination spas which are located near natural springs and offer services that incorporate mineral waters, and this is an important sub-category of destination spas. Spa industry associations define a mineral spring spa as one offering an on-site source of natural mineral, thermal or seawater, which are used in hydrotherapy treatments.

Such destination spas are becoming increasingly popular as the awareness of, and demand for, water-based therapy benefits grows. As such, more hotels and resorts are being developed that either incorporate a spa with an on -ocation mineral water source or developed in close proximity to a stand-alone mineral spa, to provide the accommodation demand created by spa tourism.

What makes a hotel spa popular?

Hotel guests have the expectation that a quality hotel will feature a spa facility. A hotel without a spa is considered less appealing in comparison to a hotel with a spa. Spa guests automatically have certain expectations of a spa, including its facilities, services, cleanliness, friendliness and approachability of staff.

However, there are a number of key factors that will differentiate guest satisfaction, and therefore the amount of guest spending and repeat business generated at the property. Hotel guests will visit the spa for a number of key reasons, and a good spa will ensure the guest's requirements are met consistently.

Type and design of facilities

The type and design of the spa facilities can play an influential role in the marketing and positioning of the hotel by improving guest perception of the property even before visiting. Well-designed and unique spa facilities will facilitate positive guest experiences that can be enjoyed for one hour, all day or in sequence through a series of days or weeks, depending on specific programmes offered and the particular selection of services.

The type and range of facilities interwoven by effective design, supported with a selection of compelling services and experiences, creates an environment where guests are eager to spend time in the spa.

Intelligently designed environments combined with unique programming and service features enable guests to schedule more time at the spa, thereby increasing their spending opportunities.[2]

Pampering and feeling special

A spa is sometimes the 'ultimate indulgence' for certain guests. As such, a spa that can enhance this feeling by making the guest 'feel special' will increase the likelihood of generating repeat business. It is easy for new spa guests to be uncertain as to what to expect upon arrival to a spa.

The importance of well-trained staff at every touch point of the guest experience, from arrival through service delivery to departure, cannot be overemphasised.

The spa is an important extension of the hospitality culture established by every department within the hotel, but by its very nature in delivering personalised services, there are more opportunities to enhance and customise guest experiences.

Quality services

The type and quality of spa services provided are paramount to creating positive guest impressions and experiences. Quality management and employee training programmes are important to enhancing the level of service in the spa. In addition, it is important that these treatments are aimed at the target market, and they communicate the offering effectively, so the market knows what is being offered.

Visitors are looking for results whether they desire relaxation or aesthetic services, yet the quality of service delivery will also significantly influence guest spending, volume of bookings and likelihood of higher retail sales.

Value for money

Guests have the need to receive value for money even at the very 'top end' of the market. The relative cost of services must reflect the quality being offered, though it must be stated that providing a discounted or inexpensive treatment because it is 'cheap' can adversely impact business success.

At the 'top end', the cost of treatments may be less important, but only as long as the quality of the experience and the 'sense of exclusivity' of the offering is exceptional.

Basic valuation methodology

As with all trading properties, the nature of each spa business is so varied that direct comparable evidence is not easy. A destination spa is unlikely to have the same business profile as a city centre hotel spa with an extensive membership business, even if the treatments and facilities are similar.

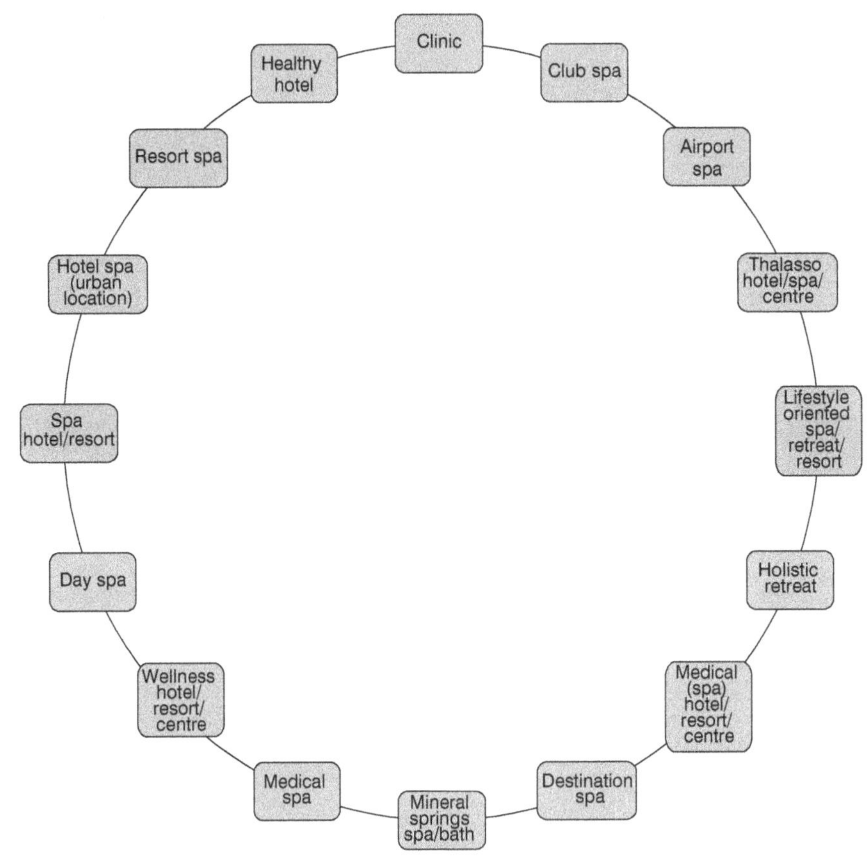

Figure 17.1 Types of spa

In addition, the lack of published spa investment and financial performance data on the open market tends to lead to a lack of comparable evidence that valuers and future developers can rely upon.

The Uniform Systems of Accounts for the Lodging Industry (USALI) categorizes the spa as an other operated department within a hotel profit and loss statement with many of its operating expenses considered undistributed operating expenses. The spa department profit and loss statement will therefore require a thorough review to ensure it is an accurate representation of all financials actually relating to the spa facilities and services. Many hotels create sub-schedule lines in the profit and loss statement for the spa department revenues and expenses which greatly simplifies the process of tracing and identifying designated revenues and expenses.

Obtaining the true financial picture of the spa may, in certain circumstances, be made more challenging because of management decisions to allocate various spa revenues and expenses to different departments in the financial reporting. For example, properties may choose not to follow the USALI and apply all

Figure 17.2 Spa relaxation area in the Chedi, Oman

undistributed operating expenses considered applicable to the spa facility directly to the spa department. In another example, the spa may not accurately be charged for energy and other costs because the property does not measure the spa energy utilization separately from the main property. In both examples, the bottom-line performance may be distorted and will need to be adjusted appropriately if the valuation is to be accurate.

A review of spa financials can be frustrating in relation to costs being difficult to assess. Increasingly, 'well-being hospitality'[3] trends are becoming more complex, with properties incorporating well-being activities, services and products throughout the hotel, including food and beverage, catering, events, rooms and recreation, restaurants with menus designed by celebrity 'healthy lifestyle' chefs or nutritionists, in-room meditation/sleeping enhancement amenities, health snack mini-bar, in-room workout equipment, personalised online streaming workouts, jogging, hiking, aquatic and bike ride programmes.

When all spa revenue and operating costs are correctly listed, the method of valuation is the profits method, although it must be stressed that comparable evidence is still important for each category of the valuation under the profits method.

The profits method brings measurability to the valuation process which is becoming more important as spas are demanding more investment, based on the complexity of the spa designs that are increasingly occupying more hotel real estate. The valuation process brings transparency in identifying the spa's contribution to the overall hotel value.

Greater spa financial transparency may also influence hotel room rate decisions, particularly if spa investment and operational costs have been poorly factored into the hotel budget.

Figure 17.3 Spa facilities in Fujairah Rotana Resort and Spa

Comparable transactions where available will help define yield selection and comparable trading profiles will help determine potential trading profitability for the property.

Example: Root Spa

The Root Spa is a minimal spa within a hotel. It offers limited spa with facilities and treatments, including massages and beauty treatments. Treatment revenue represents 72.3 per cent of turnover, with other income generating just under 11.9 per cent and product retail 15.8 per cent of all income in the fourth year of trading.

Gross operating profit equates to 19.4 per cent, which reduces to a net operating profit once fixed costs and an FF&E reserve have been deducted of between 9.6 per cent and 13.9 per cent of total revenue.

The appropriate capitalisation rate for such a spa is considered to be 12.0 per cent, resulting in a gross value of €870,000.

Operational challenges – getting to the bottom line

When undertaking the valuation of a spa, getting to the sustainable EBITDA of the spa and what this measurably contributes to the EBITDA of the property is the immediate goal.

Table 17.1 Root Spa, forecast in present values

	2016	%	2017	%	2018	%	2019	%
Revenues (€)								
Services	465,466	74.6%	518,095	73.9%	569,905	73.2%	626,895	72.3%
Other income	76,998	12.3%	85,468	12.2%	94,015	12.1%	103,416	11.9%
Retail	81,370	13.0%	97,644	13.9%	114,243	14.7%	137,092	15.8%
Total revenue	623,834	100.0%	701,206	100.0%	778,162	100.0%	867,403	100.0%
Cost of sale (€)								
Product service cost	54,246	11.7%	60,356	11.6%	66,392	11.6%	73,031	11.6%
Product retail cost	32,548	40.0%	39,057	40.0%	45,697	40.0%	54,837	40.0%
F&B								
Total cost of sale	86,794	15.9%	99,414	16.1%	112,089	16.4%	127,868	16.7%
Staff expenses – payroll	324,000	51.9%	356,400	50.8%	384,912	49.5%	404,158	46.6%
Direct operating expenses (€)								
Administration and general	31,192	5.0%	31,348	4.5%	31,504	4.0%	31,662	3.7%
Sales and marketing	43,668	7.0%	43,887	6.3%	46,081	5.9%	48,385	5.6%
Energy	34,311	5.5%	36,463	5.2%	39,686	5.1%	44,238	5.1%
Repairs and maintenance	12,477	2.0%	12,539	1.8%	31,126	4.0%	34,696	4.0%
Laundry	13,101	2.1%	14,725	2.1%	16,341	2.1%	18,215	2.1%
Others	15,596	2.5%	17,530	2.5%	19,454	2.5%	21,685	2.5%
Total direct operating expenses	119,152	19.1%	125,144	17.8%	152,689	19.6%	167,219	19.3%

Income before fixed costs (€)	93,887	15.1%	120,248	17.1%	128,472	16.5%	168,157	19.4%
Fixed costs (€)								
Reserve for renewals	15,596	2.5%	17,530	2.5%	19,454	2.5%	21,685	2.5%
Rates and insurance	18,715	3.0%	21,036	3.0%	23,345	3.0%	26,022	3.0%
Total fixed costs	34,311	5.5%	38,566	5.5%	42,799	5.5%	47,707	5.5%
EBITDA (€)	**59,576**	**9.6%**	**81,682**	**11.6%**	**85,673**	**11.0%**	**120,451**	**13.9%**

Table 17.2 Root Spa income capitalisation method (in €s)

Stabilised EBITDA in present values	120,451	
Capitalised at 12.0%	8.333	1,003,757
		1,003,757
Less income shortfall		134,421
Gross value		869,336
Net value		869,336
Net value – SAY		870,000

It is usually performed through five quite separate exercises:

1 property inspection
2 analysis of the historic trading accounts
3 review of the competition
4 interviewing the property management and relevant personnel managing the spa
5 independent analysis of everything that has been seen and said throughout the valuation fact -inding process to determine how the spa is likely to trade in the hands of a reasonably efficient operator (REO).

Assessing the likely trading profile of the spa will require a complete understanding of the trading environment in which the facility operates. It is difficult to provide a comprehensive check list of things that need to be considered, as each spa tends to be different.

However, breaking down the trading profile into sections is usually preferred, looking at the various income streams, followed by the various related operational costs. This is one of the most sensible approaches as it will usually result in a thorough analysis of the likely vagaries that can occur with spas.

Income

Typically, a spa will generate income from a number of different sources, and it is essential that the valuer reviews each of these in detail to be able to accurately project the likely income stream for the spa going forward.

These may include the following>

General facility use

Some hotel properties charge in-house guests an additional daily fee for use of the spa facilities, including the swimming pool, sauna, steam room, relaxation rooms and all associated experiences that are considered part of the spa.

Membership

Many hotel spas will offer membership to the facilities for external guests (non-hotel resident). However, it is important to identify if the membership categorisation is for the spa specifically and not a health club or recreational membership, where the spa is considered a sub-department.

Properties located in close proximity to affluent and well-populated communities can often generate significant membership revenues.

Day-visitor fees

Additional revenue streams can be generated by hotels and resorts that offer 'day use' options allowing non-registered guests access to saunas, steam rooms, swimming pools, tennis courts, golf and fitness facilities for a daily fee.

Services

This revenue stream tends to be the most significant single source of earnings. It includes income for all treatments offered at the facility that involves the exclusive use of a therapist, aesthetician, fitness instructor or other service provider professional.

Treatments will include such products as face treatments, body treatments (i.e. massage and body wraps), manicure and pedicure, and other relaxation and beautifying services. It will also include coaching/counselling and gym instructor sessions if the exercise facilities are considered part of the spa facility.

The services category will also include healthcare services offered by medical professionals, including conventional and alternative medicine. Preventive healthcare, now recognised as a mainstream form of healthcare, is reflected in the medical systems of Ayurvedic medicine and traditional Chinese medicine (TCM), along with state-of-the-art executive health screenings that may also focus on the overall emotional, mental and physical well-being of the individual.

Nutrition, detoxification and stress management are core to many programme services, while other spas may feature age-management and 'youthful ageing' treatments such as non-invasive cosmetic procedures to improve the face and body appearance.

Food and beverage

This is usually limited to a single F&B outlet located in or near the spa facilities. It will typically offer refreshments such as fresh juices, smoothies, teas, snacks and for those guests who may be booked for a series of services throughout the day, nutritious meal selections.

Retail

Spa retail can be a significant income stream for the facilities, especially when private label and/or branded products are used. Retail products can include personal care amenities, clothing, books and gift items.

Other income

Within the spa profit and loss statement there is usually a separate 'other income' line to cover all other income that is not included in specific individual revenue categories and these could be unique to each project.

It might incorporate items such as gift certificates, rental income of facilities or clothes, advertising or private hire. Some properties charge guests for the hiring of towels and robes that may also be included in this revenue category.

Other income may be used to represent spa revenues when it is used by the hotel to leverage hotel initiatives in guest marketing, programmes and services.

Costs

Costs are usually broken down into traditional departments, including the direct cost of sales, staffing costs, sales and marketing, administration and general, repair and maintenance, water charges, energy, cleaning, insurance and taxes.

Special care should be taken to ensure all operating costs are accurately accounted for, including items relating to operating costs of the entire spa facilities.

If the spa is being managed by a third-party operator on behalf of the hotel/ resort operator, then close attention needs to be paid to the financial terms of the management agreement and how they are reported in the financial statement.

It is important to review the cost base so it reflects the likely profitability of the specific spa and the property. It may not be possible, for example, to meet forecasted operating margins if the past performance shows usage is low, or the area is poorly designed and over-capitalised from an investment point of view.

Providing suitable services may require specialised staff, which can sometimes incur higher labour costs, depending on the type of services being provided to the guests.

The spa can often be used by the hotel operator as a gifting gesture to hotel guests allowing use of spa facilities and services without charge. A prime example includes VIP's/ suite guests and executive club guests receiving spa gift packages, which incur a cost without generating a direct revenue to the spa but it is to the revenue benefit of the hotel.

Inspection

A typical spa inspection will follow the general course of a hotel inspection. Initially, the valuer will request the information required, ahead of the inspection. This will include all, or some, of the following.

Legal information

- title documentation or reports on title
- ground leases or property leases
- subleases
- occupational licences or concessions
- planning consents for the site and each building on the property
- any restrictive covenants and other use prohibitions
- details of any unused planning consents
- details of any unfulfilled planning conditions
- copies of the operating licences
- agreements in place with doctors or other professionals, etc.
- management contracts
- franchise agreements.

Property information

- area plan with square metre/feet of the site
- site plans (with sizes) for each building
- building plans – floor by floor
- breakdown of spa facilities (by area)
- breakdown of other facilities, including supporting F&B outlets
- copies of any current or historic valuation reports
- copies of any current or historic condition surveys, asbestos surveys etc.
- construction cost that can be associated to the spa and its facilities
- breakdown of the original spa FF&E investment
- details of historic capital expenditure on the properties (last three years)
- details of any proposed capital expenditure (this year and next year's budget).

Operational information

- historic profit and loss accounts (three years minimum)
- management statistics, including all spa performance data such as number of visitor numbers (hotel resident and non-hotel resident), average number of services and average service cost and ticket item
- statutory accounts
- profit and loss budget for the current year and for the next year
- details of market segmentation and geographical analysis of the members and their guest profile
- records of complimentary spa entrance and gifting to guests (spa facilities and services can often be used by the hotel operator as a gifting gesture to guests without associated revenue)
- membership numbers and rates broken down by type of membership
- staff details – numbers of staff and contracts of employment, including pension details, length of service and skills training for each member of staff

- sales and marketing plans; this year and next year's budget
- details of f and b revenues; average cover spends, etc.
- details of the local markets; SWOT analysis of the local competitors
- details of local trading; how does the property compete with its competitive set?
- cross-out charges and principles of certain overhead costs that are proportionally allocated to the spa operation
- quality assurance audit records when done by either the hotel operator or by a hotel affiliation organisation or consulting body.

Other

- details of any outstanding complaints, legal claims, etc.
- details of the tax positioning of the property (outstanding tax bills, capital allowances, etc.)

The spa will be inspected, usually in the presence of the manager. The valuer will be looking at the spa facilities to see that it is suitable for use and how appealing it will be to the local or in-bound target markets.

This may relate to the quality and type of facilities, the design and number of service areas. Specific care will be taken where the spa forms part of a resort or hotel, to ensure it works effectively as a stand-alone unit, while seamlessly integrating with the larger property around it.

In addition, the valuer will be noting details about the following:

- Car parking – is it suitable for the volume of guests? Could it handle additional business without creating operational difficulties? Is there spare capacity that could be used for an alternative use?
- Non-income-generating space – are the designated spa facilities and service areas appropriate for the market size and needs?
- Rights of way – are there any public rights of way? Do they impinge on the operation of the business?
- Other facilities – are there any other areas that could generate income for the business? Are there any items that might raise costs for the business?
- Equipment – is all the equipment up to date and suitable for use? Does it have operational issues that need to be reflected in the trading profile? Will it need to be replaced soon?
- Health and safety – does the property comply with current and future legislation? If not, what are the cost implications?

The purpose of the inspection is to see all matters that may impact on revenue generation or have cost implications, so an accurate assessment of future trading can be made.

The condition and life span of the FF&E within the spa facilities will also need to be assessed to ensure that any future capital expenditure requirements can be accurately gauged.

The quality of the spa is then reviewed in relation to the relevant market, so an accurate assessment of the appropriate yield for the property can be made to finally determine value.

Interview with the general manager

After the inspection of the property, the trading history, past, present and future, will be discussed with the management.

Typically, a detailed discussion will review the operational issues relating to the spa's financial performance and the accounting principles applied in terms of revenue and cost allocations, followed by anticipated costs. This should result in a full understanding of the future trading profile and a review of the proposed capital expenditure plans.

Understanding the hotel guest utilisation levels of the spa is helpful to gain a perspective on whether the facilities make a notable contribution to the hotel operation. It provides an indication if the spa is being marketed properly to both internal and external guests. Revenue generation and profitability for the property should be discussed, bearing in mind that new supply could potentially impact on membership numbers, day visits and treatment numbers. A detailed discussion with the general manager regarding the various strengths and weaknesses of the spa, in relation to the existing (and proposed) market, is required to allow the valuer to create an understanding of the local market dynamics as seen from the perspective of the hotel/resort manager.

The number of services predicted to be provided, by type, along with overall projected average treatment costs will be discussed in light of past trading so an understanding of the projections can be fully made.

The same process will occur for each revenue category, followed by each cost line to ensure trading projections are typical of the target markets, and achievable.

The valuer will ask to review any licences that are required for operation, discuss the ownership of the property, ownership of the equipment and details of any franchises or concessions.

Example: Captain Cook Spa

In this example, the spa forms part of the overall destination with larger spa facilities including wet areas. The exercise facilities and supporting food and beverage outlet form part of the spa operation. The instruction in this instance is to calculate the value of the spa on a stand-alone basis because the owner is considering outsourcing the entire spa operation to a new management company and therefore wishes to gain a better financial overview of the spa.

There is a lack of transparency in the spa financial performance given that certain spa facilities were offered to guests without charge (and without an internal cross charge), and therefore without any associated and traceable spa revenue appearing in any property financial documentation.

Table 17.3 Captain Cook Spa: forecast in present values

	2015 (First full operational year)	%	*2016*	%	*2017*	%	*2018*	%
Revenues (€)								
Day-visitor fees	225,000	12.7%	270,000	11.5%	288,000	12.8%	315,000	12.6%
Services	800,206	45.3%	1,155,284	49.2%	974,381	43.4%	1,071,819	43.0%
Other income	76,998	4.4%	85,468	3.6%	94,015	4.2%	103,416	4.1%
Retail	163,431	9.3%	201,020	8.6%	247,255	11.0%	304,123	12.2%
Food and beverage income	198,397	11.2%	279,489	11.9%	257,715	11.5%	287,615	11.5%
Spa utilisation credit (SUC)	301,125	17.1%	355,875	15.2%	383,250	17.1%	410,625	16.5%
Total revenue	1,765,157	100.0%	2,347,136	100.0%	2,244,615	100.0%	2,492,599	100.0%
Cost of sale (€)								
Product service cost	175,441	21.9%	241,947	20.9%	202,995	20.8%	217,418	20.3%
Product retail cost	65,372	40.0%	71,910	35.8%	79,101	32.0%	87,011	28.6%
Food and beverage cost	79,359	40.0%	111,796	40.0%	103,086	40.0%	115,046	40.0%
Total cost of sale	320,172	27.6%	425,652	26.0%	385,182	26.0%	419,475	25.2%
Staff expenses – payroll (€)	673,200	38.1%	706,860	30.1%	763,409	34.0%	801,579	32.2%

Direct operating expenses (€)								
Administration and general	52,032	2.9%	52,292	2.2%	52,553	2.3%	52,816	2.1%
Sales and marketing	72,844	4.1%	73,209	3.1%	76,869	3.4%	80,713	3.2%
Energy	349,746	19.8%	358,489	15.3%	367,452	16.4%	376,638	15.1%
Repairs and maintenance	70,606	4.0%	93,885	4.0%	89,785	4.0%	99,704	4.0%
Laundry	51,190	2.9%	68,067	2.9%	65,094	2.9%	72,285	2.9%
Others	81,197	4.6%	71,522	3.0%	70,215	3.1%	77,865	3.1%
Total direct operating expenses	625,583	35.4%	717,464	28.3%	669,414	29.8%	707,205	28.4%
Income before fixed costs (€)	146,201	8.3%	549,451	23.4%	426,610	19.0%	564,339	22.6%
Fixed costs (€)								
Reserve for renewals	44,129	2.5%	58,678	2.5%	56,115	2.5%	62,315	2.5%
Rates and insurance	52,955	3.0%	70,414	3.0%	67,338	3.0%	74,778	3.0%
Total fixed costs	97,084	5.5%	129,092	5.5%	123,454	5.5%	137,093	5.5%
EBITDA (€)	**49,118**	**2.8%**	**420,359**	**17.9%**	**303,157**	**13.5%**	**427,246**	**17.1%**

Furthermore, many spa undistributed operating expenses appeared in the spa as operating expenses. These expenses were costs traceable to benefits by generating a revenue for the hotel.

As such, to create an equilibrium in the valuation, an element of income from the hotel rooms has been included in the spa revenue as an allowance for hotel guest use of the spa facilities (sauna, steam room, swimming pool, experience pools, exercise facilities, etc.) known as a spa utilisation credit (SUC). The SUC is calculated on a per occupied room basis and serves as a transparent internal accounting procedure to account for spa facilities that are used by hotel guests and have no associated direct revenue.

These facilities do have direct operating costs such as energy, laundry, manpower, repair and maintenance that require consideration. The hotel promotes the spa facility as a hotel amenity that guests can use without additional charge to the published room rate.

In all cases where the SUC is applied, the SUC fees would need to be investigated to ensure they are realistic. There is usually much discussion and possible disagreement between the owner and hotel operator regarding the application of SUC due to the differentiating financial objectives of each party. The SUC concept creates the parameters for gaining a better understanding of the spa contribution to the overall property financial performance by establishing all measurable revenue and expense allocation.

In this example of a start-up spa, we see expanded revenue sources creating a wider income spread. Treatment revenue is the main source of income where it contributes just under 42 per cent of turnover, day-visitor fees at 12.3 per cent, food and beverage 11.2 per cent and retail 11.9 per cent in the fourth year of trading. A revenue adjustment is made with the SUC revenue line which equates to 18.7 per cent of total revenue. The noticeable contribution of SUC can be put into its rightful perspective when considering that many of the direct operating expenses related to operating the spa will be a result of serving hotel guests

The appropriate capitalisation rate for such a property is deemed to be 9 per cent, resulting in a gross value of €4,250,000.

Table 17.4 Captain Cook Spa income capitalisation method (in €s)

Stabilised EBITDA in present values	427,246	
Capitalised at 9.0%	11.111	4,747,183
		4,747,183
Less income shortfall		509,105
Gross value		4,238,077
Net value		4,238,077
Net value – SAY		4,250,000

Stand-alone spa

There are many types of stand-alone spa, all of which have different facilities, operations and investment criteria. This includes beauty spas, day spas, health club spas, medical spas, mineral spring spas and many more that will further vary, depending on the geographical location.

The common denominator in a stand-alone spa will be that it is not part of a hotel operation and operates without providing overnight accommodation. Therefore, they will be valued in a slightly different manner than a spa in a hotel. More stand-alone spas are being integrated into mixed-use developments that also contain a hotel, residential units and entertainment facilities. In such cases, the stand-alone spa is not be built within the hotel, and is operated as a completely separate entity to the hotel. Hotel guests may have access to the stand-alone spa facilities and services but utilisation will be either cross charged from the spa to the hotel based upon a guest fee agreement between the stand-alone spa and hotel, or the guest will be charged directly for spa access by the spa operator.

The real estate value of a stand-alone spa may have higher alternative use value as compared to a hotel spa and therefore factors significantly in the valuation method applied along with the profit method. For example, a day spa operating in a prime city centre location where the operator also owns the real estate will have to consider the planning environment that affects the property to assess potential alternative uses.

The real estate could potentially have many alternative uses, some of which may be more valuable than its use as a day spa.

In the case of a stand-alone mineral spa, the land asset will include the potential ownership and/or licensing of the mineral water to supply, which could enhance value.

Given the various types of stand-alone spas, it is once again important to understand that different types of stand-alone spa properties will have differentiating spa facilities with operational variations that clearly need to be understood.

Example: Flintoff's

A stand-alone spa's financials will conform to much more of a traditional financial reporting without the ambiguity that we sometimes see in a spa located in a hotel. In this example, the property is trading well and is well-suited to the market's needs, and as such, the existing use value is higher than an alternative use value.

This property is a stand-alone thermal spa with extensive facilities, catering to thermal bathing and related services. It has just been completely refurbished after having been closed for eighteen months.

Such spas typically are more complex operating businesses given the number of ancillary product lines, including F&B outlets and retail outlets that are operated by a third party on a concession agreement.

Table 17.5 Flintoff's Spa: forecast in present values

	2016	*%*	*2017*	*%*	*2018*	*%*	*2019*	*%*
Revenues ($)								
Day-visitor fees	3,418,538	46.1%	4,045,642	44.2%	4,322,359	41.1%	4,622,359	40.8%
Services	2,349,245	31.7%	3,005,347	32.9%	3,523,500	33.5%	3,786,291	33.4%
Other income	216,866	2.9%	266,747	2.9%	299,947	2.9%	320,755	2.8%
Retail	399,372	5.4%	469,849	5.1%	661,176	6.3%	813,247	7.2%
Food and beverage income	428,388	5.8%	570,344	6.2%	724,054	6.9%	755,773	6.7%
Membership	598,500	8.1%	785,531	8.6%	989,769	9.4%	1,039,258	9.2%
Total revenue	7,410,909	100.0%	9,143,460	100.0%	10,520,805	100.0%	11,337,682	100.0%
Cost of sale ($)								
Product service cost	469,849	20.0%	601,069	20.0%	704,700	20.0%	757,258	20.0%
Product retail cost	159,749	40.0%	187,940	40.0%	264,471	40.0%	325,299	40.0%
Food and beverage cost	171,355	40.0%	228,138	40.0%	289,621	40.0%	302,309	40.0%
Total cost of sale	800,953	25.2%	1,017,147	25.1%	1,258,792	25.6%	1,384,866	25.9%
Staff expenses – payroll ($)	3,081,154	41.6%	3,389,269	37.1%	3,660,410	34.8%	3,825,129	33.7%
Direct operating expenses ($)								
Administration and general	148,218	2.0%	163,040	1.8%	177,714	1.7%	193,708	1.7%
Sales and marketing	370,545	5.0%	457,173	5.0%	468,176	4.5%	515,865	4.6%
Energy	518,764	7.0%	548,608	6.0%	568,123	5.4%	612,235	5.4%
Repairs and maintenance	185,273	2.5%	228,587	2.5%	263,020	2.5%	283,442	2.5%

Laundry	176,380	2.4%	217,614	2.4%	250,395	2.4%	269,837	2.4%
Others	340,902	4.6%	365,738	4.0%	362,968	3.5%	396,819	3.5%
Total direct operating expenses	1,591,863	21.5%	1,817,720	19.9%	1,912,682	18.2%	2,078,197	18.3%
Income before fixed costs ($)	1,936,939	26.1%	2,919,324	31.9%	3,688,920	35.1%	4,049,490	35.7%
Fixed costs ($)								
Reserve for renewals	185,273	2.5%	228,587	2.5%	263,020	2.5%	283,442	2.5%
Rates and insurance	222,327	3.0%	274,304	3.0%	315,624	3.0%	340,130	3.0%
Total fixed costs	407,600	5.5%	502,890	5.5%	578,644	5.5%	623,573	5.5%
EBITDA ($)	**1,529,339**	**20.6%**	**2,416,434**	**26.4%**	**3,110,276**	**29.6%**	**3,425,917**	**30.2%**

Table 17.6 Flintoff's Spa income capitalisation method (in $s)

Stabilised EBITDA in present values	340,130	
Capitalised at 8.75%	11.429	3,887,205
		3,887,205
Less income shortfall		$208,136
Gross value		3,679,069
Net value		3,679,069
Net value – SAY		3,675,000

Revenue sources for a stand-alone thermal spa are greatly influenced by tourism and local community usage. As such, day-visitor fees and services make up 74.2 per cent of the income in the fourth year of trading. Given the high footfall to such spas retail can potentially be very high, while other income lines can seem high if the spa has retail tenants who rent space and if space is available for event hosting.

The appropriate capitalisation rate for such a property is 8.75 per cent, resulting in a value of $3,675,000.

Notes

1 Statista.
2 Ferrari, S., Puczkò,L. and Smith, M. (2014). 'Co-creating Spa Guest Experience'. In Kandampully, J., ed., *Guest Experience Management: Enhancing Experience and Value through Service Management.* Dubuque, IA: Kendall Hunt.
3 Well-being hospitality is an overarching approach to improving health, happiness and contentment within hospitality and destination assets via forms of wellness, leisure, recreation, travel and healthcare.

Glossary

à la carte menu a food and beverage menu in which each item is listed and priced separately.

A&G administration and general.

adjoining rooms Guestrooms located side by side without a connecting door between them.

ADR (alternative dispute resolution) the range of processes for resolving commercial disputes without seeking redress from the courts. In addition to arbitration, this most commonly means mediation, adjudication and conciliation. Some pre-litigation protocols require, or strongly encourage, the use of ADR – particularly mediation – prior to commencing legal proceedings, with potential consequences in the award of costs against non-complying parties.

ADR (average daily rate) calculated by dividing actual daily revenue by the total number of rooms sold. affiliated hotel A hotel that is a member of a chain, franchise, or referral system. Membership provides special advantages, in particular, a national reservation system.

ADS (alternate distribution system) *see* IDS.

AGA (authorised guarantee agreement) the concept was introduced by the Landlord & Tenant (Covenants) Act 1995 to implement changes to the law on Privity of Contract on Assignment of a commercial property lease. It is an agreement that an outgoing tenant enters into with the landlord when it assigns its lease to a new tenant. Under the AGA, the outgoing tenant guarantees the performance of the covenants by the new tenant. The outgoing tenant therefore becomes the guarantor for the new tenant.

AGOP (adjusted gross operating profit) equal to the gross operating profit minus the hotel management base fee and any additional expenses.

air handling unit An all-air HVAC system consisting of coils (through which steam/hot water or chilled water is circulated from central boilers and chillers), filters, fresh air intakes, exhaust air dischargers, and sometimes humidification equipment.

alienation the legal transfer of title of ownership to another party.

alienation provision clause(s) in a lease which govern a tenant's ability to dispose of their leasehold interest.

ALOS (average length of stay) calculated by dividing the number of room nights by the number of bookings.

alterations works undertaken by the tenant which alter, cut, divide, relocate or modify the demise, which can include the building, partitions, mechanical and electrical services or finishes.

arbitration is governed by statute; agreements to refer disputes to a specialist arbitrator are often made in a lease or building contract. Arbitration is private; the arbitrator's award is final and binding and is based on evidence put forward. There are limited rights of appeal to the courts on procedural irregularities and points of law. This is a popular way of resolving property disputes where privacy and speed are important.

assignment an assignment of a lease is where the tenant transfers/sells their entire interest in the property, for the unexpired term of the lease, to an assignee.

attrition rates the level of membership lost per year; calculated by dividing the membership base by the number of lost members

average food check a calculation that works out the average food-spend per visit to the restaurant or bar; it can be calculated in many ways, and with or without drinks, but is in essence a tool to track typical customer spending patterns.

B&B (bed and breakfast) room rates that include bed and breakfast.

BAR (best available rates) typically rates that are the 'best available' at the time of booking, often these are short lead bookings.

BOH (back of house) the functional areas of a hotel or restaurant in which employees have little or no direct guest contact, such as kitchen areas, engineering and maintenance, and the accounting department.

booking engine an online system used by hotels that allows prospective hotel guests to check availability and make reservations at the hotel.

break clause (alternatively called a 'break option' or 'option to determine') is a clause in a lease which provides the landlord or tenant with a right to terminate the lease before its contractual expiry date, if certain criteria are met.

break notice the formal notification that one party wishes to exercise its right to terminate the lease ('a break clause', 'option to determine' or 'break option'). Break notices must be served correctly and require a degree of care to ensure the right is successfully exercised. For example, the notice must be served the correct number of months before the break date and may require compliance with pre-conditions.

break option *see* break clause.

BREEAM (Building Research Establishment Environmental Assessment Method) a recognised environmental assessment method and rating system for buildings, which was first launched in 1990. BREEAM is one of the most commonly used standards for best practice in sustainable building design, construction and operation and is a widely recognised measure of a building›s environmental performance. A BREEAM assessment evaluates a building's specification, design, construction and use, such as energy and water use, the internal environment (health and well-being), pollution, transport, materials, waste, ecology and management processes.

BRG (best rate guarantee) the promise that hotels or OTAs will offer the best rates on their own site as compared to any other side for the same product.

buffet service hot and cold foods attractively arranged on platters are placed on large serving tables and guests walk up to help themselves. Sometimes each course is placed on a separate table. Service personnel, such as carvers, may be required to assist guests.

building insurance covers the landlord against damage, destruction and loss of rent; the cost of the insurance premium is normally reimbursed by the tenant.

building survey a report on a building by a building surveyor on the structural integrity of a building and its state of repair. This report would usually cover the condition of the structure, incidence of any defects to the fabric of the building and the state of repair of fixtures and fittings, services and plant installations. A building survey is often required by a lender prior to agreeing to a loan which is secured against the value of the property in question.

business mix a hotel's desired blend of business from various segments such as business transient, corporate group, leisure, and convention.

C&B conference and banqueting.

C&E conference and events.

C1 UK use-classes order that contains hotels, boarding and guest-houses where no significant element of care is provided (excludes hostels). There are no permitted changes.

Calderbank offer in commercial property: a written offer to settle, frequently in relation to a rent review made 'without prejudice' save as to costs. Either landlord or tenant can make a Calderbank offer. A Calderbank offer may only be referred to after the arbitrator, tribunal or court has made their decision on the substantive matter. The intention of a Calderbank offer is to protect a party against costs.

Capex (capital expenditure) funds used by an organisation to acquire or upgrade physical assets, such as property or equipment.

cashflow cash receipts minus cash payments over a defined period.

caveat emptor buyer beware.

CDP (chef de partie) in charge of a particular area of production within the kitchen.

central reservation office part of an affiliate reservation network; a central reservation office typically deals directly with the public, advertises a central (usually toll-free) telephone number, provides participating properties with necessary communications equipment, and bills properties for handling their reservations.

chain operating management a firm that operates several properties, such as Holiday Inn Worldwide or Hilton Hotels Corporation; such an operator provides both a trademark and a reservation system as an integral part of the management of its managed properties.

change of use the ability to change the way in which land or buildings are used, either by a simple alteration in the nature of the use, or through alterations and additions which modify the use. A change of use may also arise through a material intensification in the present use, or by subtly altering the present use to a point where the changes amount to development. There are certain types of change of use which do not require planning permission; for example, a change of use from one type of shop to another. Certain changes are permissible between and within the use-classes without the need for planning permission, subject to satisfying the appropriate criteria. Other uses are considered *sui generis*; that is, they are uses on their own, unrelated to other uses. A change of use from, say, a field to a caravan park would require planning permission. The permitted development rights and then the use-classes order should first be considered to determine whether an intended change of use requires planning permission or not.

channel management the application used to control the allocation of hotel inventory and rates across all distribution channels including website, third parties, and the GDS.

chef de partie *see* CDP.

chef de rang in French service, the employee responsible for taking orders, serving drinks, preparing food at the table, and collecting sales income. If there is no sommelier or wine steward, the chef de rang may serve wine.

clawback (also known as 'overage') is a right to receive future payments which are triggered by future events – for example, achieving planning permission for change of use or development, practical completion of a development, or the sale or lease of the completed development.

commis chef a basic chef in larger kitchens who may have just completed training or still be in training.

commonhold a system of freehold ownership of units suitable for interdependent buildings such as blocks of flats based on Australian strata title.

complimentary occupancy percentage a ratio that shows the percentage of occupied rooms that are complimentary and generate no revenue; calculated by dividing complimentary rooms for a period by total available rooms for the same period. Sometimes referred to simply as complimentary occupancy.

complimentary room a complimentary or 'comp' room is an occupied room for which the guest is not charged. A hotel may offer comp rooms to a group in ratio to the total number of rooms the group occupies. For example, one comp room may be offered for each fifty rooms occupied.

compulsory purchase an acquisition of interests in land or rights, generally by a public body such as a government department or local authority, which is authorised by an appropriate compulsory purchase order (CPO). This process entitles the purchaser to acquire from an unwilling owner or occupier. This is most commonly related to proposals for future public works such as transport developments and can come from local authorities, highways or government agencies. Anyone who has land acquired as part of a CPO is generally entitled to compensation. There are a number of rules and regulations governing the assessment of compensation, which not only relate to the value of the land and buildings, but also losses arising from disturbance, severance and injurious affection, and in circumstances where no land is taken.

concierge an employee whose basic task is to serve as the guest's liaison with hotel and non-hotel attractions, facilities, services, and activities.

condition precedent a lease covenant including conditions which must be strictly fulfilled to satisfy the requirements of the lease.

connecting rooms two or more guestrooms with private connecting doors permitting guests access between rooms without their having to go into the corridor.

contaminated land land that is contaminated contains substances in or under the land that are actually or potentially hazardous to health or the environment. Britain has a long history of industrial production and throughout the UK there are numerous sites where land has become contaminated by human activities such as mining, industry, chemical and oil spills and waste disposal. Contamination can also occur naturally as result of the geology of the area, or through agricultural use.

continental breakfast a small morning meal that usually includes a beverage, rolls, butter, and jam or marmalade.

contracting out commercial property leases generally automatically qualify for the protection afforded to tenants at lease expiry by the Landlord & Tenant Act 1954. The parties to a lease may, by agreement, contract out of the Act. The main consequence of so doing is to remove the tenant's rights of renewal, and eligibility for compensation in certain circumstances.

corporate rates rates negotiated by corporates/companies with a hotel or sales team. Typically, these are lower than standard consumer rates as corporates/companies can offer a high annual volume of bookings.

COS cost of sale.

cover diners within a restaurant; for example, a hotel restaurant may achieve 30 covers, meaning 30 people dined.

CPI (consumer price index) a measure that examines the weighted average of prices of a basket of consumer goods and services. The CPI is used to track price changes associated with the cost of living. In the commercial property market leases are often granted with rent reviews occurring by reference to either the consumer price index, or more commonly, the retail price index (RPI) (normally on an upwards-only basis). There are a number of differences between the CPI and RPI indices – the most well-known of which is in the area of mortgage payments, which are excluded from the CPI but included in the RPI. The CPI was first introduced in 1996 and in 2003 the government announced that the UK inflation target would be based on the CPI, replacing the retail price index for this purpose.

CRM (customer relationship management) commonly automated to include pre- and post-stay elements, along with loyalty programmes etc.

CRS (central reservation system) an application used to manage a hotel's distribution and hotel room bookings. Typically, the CRS will be used to reach guests via multiple distribution channels such as travel agencies, online travel agencies (e.g. Expedia, Orbitz, Travelocity, Priceline and others), direct to the hotel website, and telephone (either via call centre, direct to property or both).

CV cardiovascular.

D&B design and build

dataroom an area where all due diligence information is stored for property transactions. These tend to be electronic rather than physical 'rooms'.

day guests guests that arrive and depart the same day.

DBB (dinner, bed and breakfast) rates that include dinner, bed and breakfast.

DCF (discounted cashflow) an analysis technique used to appraise, for example, investment and development projects, whereby future income streams are discounted, accounting for the time cost of money, to arrive at a present value in order to gauge performance and project viability.

deed a special legal document that is clearly intended to be a deed and is signed, witnessed, attested and delivered.

demand-based pricing applying strategies to move rates based upon demand within the marketplace and what the market will bear.

demand generators strategies or programmes (social media, promos) that attract or drive demand to a specific hotel.

demise demised premises are the extent of the premises included within a lease and may include land or other facilities; for example, when an office block is let under a written lease, in the lease the office block might be referred to as the demised premises.

depreciation a method of allocating the cost of a tangible asset over its useful life. Businesses depreciate long-term assets for both tax and accounting purposes. Consumption includes the wearing out, using up or other reduction in the useful economic life of a tangible fixed asset whether from use, passing of time or obsolescence through either changes in technology or demand for goods and services produced by the asset.

dilapidations breaches of a tenant's lease covenants in respect of repair, reinstatement of alterations, and redecoration. Can be raised by a landlord during the term of the lease (interim dilapidations) or, more commonly, at lease expiry (terminal dilapidations). Any resultant claim for damages is capped at the Diminution in Value of the landlord's interest under section 18 of the Landlord and Tenant Act 1927 in respect of repair, and similar principles at common law in respect of reinstatement and redecoration.

displacement analysis analysis of group business based upon the total value of the business versus what transient business would be displaced if the business were accepted. The group value includes all food and beverage spending, meeting room rental and any additional outlet spending minus any costs involved.

distribution strategy determining when and through which channels to sell rooms based upon the cost of acquisition of the individual channel. Hotels can maximize their profitability by looking at total costs associated to each strategy.

doorknob menu a type of room service menu that a housekeeper can leave in the guestroom. A doorknob menu lists a limited number of breakfast items and times of the day that the meal can be served. Guests select what they want to eat and the time they want the food delivered, then hang the menu outside the door on the doorknob. The menus are collected and the orders are prepared and sent to the rooms at the indicated times.

double-loaded slab a guestroom floor configuration in which rooms are laid out on both sides of a central corridor.

easement the right of the owner of one piece of land (the dominant tenement) to a benefit from other land (the servient tenement).

ecotourism low-impact tourism that avoids harming the natural or normal environment. In this relatively new approach to promoting enjoyment, as well as protection, of the environment, tourists seek out environmentally sensitive travel and/or tours or vacations which, in some way, improve or add to their knowledge of an environment.

ED (electronic distribution) encompasses all the electronic channels of distribution, which includes GDS, online travel agencies and web booking engines. These distribution channels can be accessed through the Internet, an intranet or through an interfaced connection.

EPC (energy performance certificates) are required for buildings when they are sold, built or let in the UK. The certificate identifies how energy efficient a building is by providing a rating from G (least efficient) to A (most efficient). It is accompanied by a report providing recommendations for potential improvements to the building and indicative costs, pay back periods and carbon impacts.

estoppel the legal principle whereby one party is held to have varied its rights by its actions, and in doing so has given another party sufficient encouragement to act to its detriment in a way contrary to any existing relationship.

EUV existing use value.

executive floor a floor of a hotel that offers exceptional service to business and other travellers. Also called a business floor or the tower concept.

exit yield the yield that is applied to the projected income on the assumed sale date of the investment.

expert determination expert determination in the UK involves an independent third party, acting as an expert deciding a dispute using their own knowledge and experience. This is a common way of resolving rent review and valuation disputes. It is seen as quicker and cheaper than arbitration and particularly suitable to specialist property disputes.

expert witness an expert witness is required to assist the court or tribunal to understand complex technical matters on which their decision might be based. Experience in the subject matter of the dispute is critical and the choice of expert witness can make or break a case. Expert witnesses appear in most property disputes, whether before the courts or arbitrators. While appointed by one of the parties, they have an overriding obligation of impartiality to the court or tribunal.

external valuer a valuer who, together with any associates, has no material links with the client, an agent acting on behalf of the client, or the subject of the assignment.

F&B food and beverage.

fair market share a hotel's individual percentage of market share which they should reasonably expect to capture, all things being equal and based upon their competitive set. Calculated by dividing the number of rooms at the hotel by the total number of rooms in the competitive set (inclusive of the subject hotel).

fam tour/trip a reduced-rate, often complimentary, trip or tour offered to travel agents, wholesalers, incentive travel planners, travel writers, broadcasters, or photographers to promote a hotel or a destination.

FC financial controller

FF&E furniture, furnishings and equipment.

FIDIC Fédération Internationale des Ingénieurs Conseils (International Federation of Consulting Engineers).

FIT free and independent traveller, or free independent tourist.

fit-out costs are usually incurred by a tenant prior to being able to occupy new accommodation. Fit-outs will often include everything from installing cabling through to purchasing furniture.

FOH (front of house) the functional areas of a hotel or restaurant in which employees have extensive guest contact, such as the front desk (in hotels) and the dining room(s).

food and beverage division the division in a hospitality organization that is responsible for preparing and serving food and beverages within the organization or property. Also includes catering and room service.

forfeiture when a business tenant is in rent arrears, or is in serious breach of the lease terms, then the commercial landlord will in most cases have the right to forfeit – the right to summarily end the tenancy. The landlord must, however, comply with section 146(1) of the Law of Property Act 1925. There is no automatic right to forfeit a lease unless the lease contains specific provisions by way of a clause setting out the grounds on which the landlord may forfeit. The landlord's actions must indicate that he intends to end the lease, so actions to the contrary, like accepting rent, will remove the right to forfeit. There are two main methods of doing this: (1) by peaceable re-entry to the premises or (2) by issuing court proceedings for possession.

FRI (full repairing and insuring) a term used to describe a lease where the tenant is responsible for all repairs and for insuring. However, the term also applies to the liability for payment of these costs, known as effective FRI. FRI leases therefore include those where the landlord pays for external repairs and recovers the cost via service charge or contribution to 'shared' expenditure. Also where, as is most common, the landlord maintains the insurance and recovers the cost of the premium from the tenant, usually as further rent.

front office a hotel's command post for processing reservations, registering guests, settling guest accounts, and checking guests in and out.

full board rate that includes bed, breakfast, lunch and dinner.

fully fitted a specification for a property being leased to a tenant; fully fitted is completely furnished and ready for operation.

GAAP generally accepted accounting principles.

GCC (general conditions of contract) a form of construction contract.

GDS (global distribution system) Sabre, Galileo, Amadeus and Worldspan offer a comprehensive travel shopping and reservation platform to travel agents worldwide. Agents use one of these systems to book airline, car, hotel and other travel arrangements for their customers. OTAs also use one or more GDS to power some or all of their content on their site.

GDV gross development value.

GIA (gross internal area) the internal floor area of a building measured to the internal face of the external walls; it is most commonly used in the industrial /warehouse sector, but also in food stores and retail warehousing.

GM (general manager) the chief operating officer of a hotel or a restaurant.

GOP (gross operating profit) total revenue less operating expenses.

GOPPAR (gross operating profit per available room) calculated as GOP divided by total rooms available

GOR gross operating revenue.

green hotels generally, refers to hotels making an active effort to operate sustainably and reduce their environmental impact.

gross income the total current income receivable from a property investment before allowing for any deductions.

group rates negotiated rates (usually discounted against standard rates) for group travel. This can include guests attending conferences, meetings and tours etc.

group reservations a block of multiple guestrooms that are being held under an individual or business's name at a particular hotel for a specific date or range of dates. Generally used for conventions, conferences, meetings, receptions, weddings, etc.

half board rate that includes bed, breakfast and either lunch or dinner.

head chef in charge of the kitchen, including sous chef(s), chef de partie(s) and commis chef(s).

headline rent a headline rent is the rent that is paid under a lease, after the end of any rent-free periods or any period of reduced rent. It creates an artificially inflated rent by ignoring the rent-free period, period of reduced rent or any other concessions the landlord may have given to the tenant in return for a higher headline rate. Headline rent is most commonly associated with open market lettings, but increasingly at lease renewal and lease re-gearing/ restructuring. Headline rent usually also forms the benchmark for any "upward only" rent review in the lease.

heads of terms a heads of terms agreement identifies and highlights the requirements of both the transacting parties in a property deal. Its advantage is that both parties will fully understand what they are subject to, and reduce or abolish any misunderstandings from each party. The heads of terms will form the basis of the contract and be forwarded to the parties' solicitors to draft the contract or lease.

HMA hotel management agreement or management contract.

housekeeping department a department of the rooms division, responsible for cleaning the hotel's guestrooms and public areas.

hurdle rate of return the target return or IRR from an investment.

IBE (internet booking engine) same as web booking engine (WBE).

IDS (internet distribution system) the internet and other non-GDS channels of hotel electronic distribution. Includes the internet, world wide web, intranets, extranets and online services. Also known as 'ADS' (alternate distribution system).

IHRSA International Health, Racquet & Sportsclub Association.

incentive travel travel financed by a business as an employee incentive.

inclusive tour a tour in which specific elements – air fare, hotels, transfers, etc. – are included for a flat rate. An inclusive tour rate does not necessarily cover all costs.

independent expert determination *see* expert determination.

indexation the practice of linking the review of the tenant's payments under the lease to a published index, most commonly the retail price index (RPI) but also the consumer price index (CPI). Most commonly associated with service charge payments, annual adjustment of service charge caps and rent reviews.

in-room beverage service system a computer-based system capable of monitoring sales transactions and determining inventory replenishment quantities. Two popular in-room beverage service systems are non-automated honour bars and microprocessor-based vending machines.

inventory (relative to hotel distribution) the rooms available that the hotel has to distribute/sell across all channels.

IRL (internal repairing lease) A lease where the landlord retains responsibility for, and bears the cost of, external repairs. Differs from the more common full repairing and insuring (FRI) lease, and commonly results in a higher rent reflecting the tenant's lower obligations in respect of annual expenditure on repairs.

IRR (internal rate of return) the rate of interest at which all future cashflows must be discounted in order that the net present value of those cashflows, including the initial investment, should be equal to zero.

ISPA International Spa Association.

KPI (key performance indicator) a target against which success can be measured. For example, occupancy rates, ADR or RevPAR.

last room availability strategy to allow travel sources to book the last room a hotel has available at a contracted rate.

lead time the length of time between when a booking is made and the actual stay date. Typically, hotels prefer long lead times as it allows them to plan room inventories/rates.

lease a legally binding contract between a landlord and a tenant which sets out the basis on which the tenant is permitted to occupy a property.

legal costs lawyer's costs incurred by both a purchaser and vendor of a property, or in connection with the sale or grant of a new lease by a landlord and a tenant. Such costs can include conveyancing fees, litigation advice, counsel's opinion, stamp duty, local authority search fees, bank transfer fees and court expenses, plus other disbursements and VAT.

lessee the legal term for 'tenant'.

lessor the legal term for 'landlord'.

licence an authority to do something that would otherwise be inoperative, wrongful or illegal. This may be used, for example, to permit occupation of land and buildings or allow a tenant to carry out alterations, or to assign or sublet.

liquidity the ability to convert an asset into cash within a required period.

LTV (loan to value ratio) the loan amount, expressed as a percentage of a property's market value.

LOS (length of stay) the duration of a guest's visit.

loyalty programme a rewards programme for those that stay at the hotel regularly: rewards can vary, but typically include free stays, dining vouchers etc.

M&IT meetings and incentive travel.

management agreement *see* management contract.

management contract (or management agreement) an agreement between and owner and an operator to run a hotel or resort. The agreement is quite different from a lease, and is unique to the hotel industry. The 'manager' runs the hotel on behalf of the owner.

market value the estimated amount for which an asset or liability should exchange on the valuation date between a willing buyer and a willing seller in an arm's-length transaction after proper marketing and where the parties had each acted knowledgeably, prudently and without compulsion.

market value with existing use the definition of market value with existing use disregards potential alternative uses and any other characteristics or development.

mass tourism wide-scale travel by a large number of people – not just the elite – brought about by the increase in leisure time, discretionary income, and reliable and inexpensive modes of transportation such as the automobile and airplane.

maximum length of stay a room inventory strategy that limits the number of nights a reservation can stay when arriving on a certain date.

MCI meetings, conventions and incentives.

mediation an informal method of resolving disputes in which an independent mediator works with the parties to help them resolve their dispute. The decision they reach is documented formally and then becomes binding upon them. Mediation is encouraged by the courts and has a very high success rate.

It is particularly suitable to landlord and tenant, boundary and other disputes where the parties have an ongoing relationship to preserve.

MICE meetings, incentives, conventions and exhibitions.

mini-bar a small, under-the-table unit that can be stocked with liquor, beer, and wine, usually located within a hotel room for the convenience of guests.

minimum length of stay a room inventory strategy that requires a reservation to meet or exceed a certain length of stay (e.g. two nights or more) in order to complete a reservation.

MLOS minimum length of stay.

mortgage an interest in land created as security for a debt; the lender is the mortgagee, the borrower is the mortgagor.

MPI market penetration index.

MR (market rent) the estimated amount for which a property would be leased on the valuation date between a willing lessor and a willing lessee on appropriate lease terms in an arm's length transaction, after proper marketing and where the parties had each acted knowledgably, prudently and without compulsion.

MV *see* market value.

mystery guest a quality-control measure whereby an undercover employee (usually of an external organization) poses as a guest to evaluate the performance of a hotel.

NEC 3 (New Engineering Contract 3) This is the latest updated version of a different form of contract for construction of hotels. Similar to FIDIC but it is its own form, more popularly used in the UK.

net effective rent the equivalent rent that would be payable after all incentives (for example capital contributions and rent-free periods) are taken into consideration. This calculation is used by the VOA to determine the actual rateable value. It is also used in lease negotiations to identify the appropriate level of rent.

net income the net income from a property investment after deducting ground rent and non-recoverable expenditure.

net internal area the 'useable' measured internal floor area of a building, which is measured to the internal face of external walls, by excluding 'non-useable' but ancillary/essential areas such as stairwells, WCs and permanent essential access routes. Most commonly used in the office and retail sectors.

net present value the sum of the discounted cashflow of a project, with all tranches of net income discounted to a present value at a rate derived from the investor's target rate of return or the cost of capital.

net yield Net yield takes the assumed or actual costs associated with purchasing the property into account, providing a more accurate position in respect of the relationship between the rental income and the total capital investment.

NIY (net initial yield) the initial net income at the date of purchase, expressed as a percentage of the gross purchase price including the costs of purchase.

NOP (net operating profit) sometimes referred to as EBITDA, and should (if calculated in line with USALI) provide a standard report on profitability of the property.

occ (or occupancy) the rate of occupation of a hotels total rooms, at any given time. For example, an occupancy rate of 95 per cent would mean that 95 per cent of a hotel's room inventory is presently occupied.

occupancy report a report prepared each night by a front desk agent that lists rooms occupied that night and also lists those guests expected to check out the following day.

onerous lease provisions some of the tenant's covenants in a lease may impose restrictions on their occupation of the premises which, when assessing the rental value of the property at rent review, warrant an adjustment in the rent. If the premises were offered in the open market on the same tenancy terms (and a discount could be expected in any open market bids), to reflect the onerous lease provisions in comparison to the 'market norm' for the class of property concerned, a similar discount may be applied to the rent review.

online reservation system An internet-based system used by hotels that allows prospective hotel guests to check availability and make reservations at the hotel.

open market rent the most common basis of valuation at rent review (also known as open market rental value – OMRV). It is commonly defined as the rent at which the premises might reasonably be expected to let, in the open market, at the review date, on the terms of the hypothetical lease. Typically, this is framed with primary reference to the terms and covenants of the actual lease of the premises, although this is not always the case (e.g. in relation to the assumed length of term) and due diligence is essential in understanding the particular definition of open market rent in any review clause.

open market rent review a rent review to open market rent.

optimised trading a specific contract term sometimes included in management contracts. It is slightly different from 'maximised trading' in that it balances the long-term interest of the property with short-term revenue generation.

OTA (online travel agencies) websites offering comprehensive travel shopping and reservations solutions to consumers. Examples include Expedia, Orbitz, HostelWorld, Booking, and many local and regional sites.

overage *see* clawback.

overbooking practice by hotels of confirming reservations beyond their capacity (100 per cent occupancy) in expectation of cancellations, no-shows, or in error.

P&L profit and loss account,

package tour a tour put together by a tour packager or operator. Travellers who buy the package make the trips by themselves rather than with a large group. The package offers, at an inclusive price, several travel elements which a traveller would otherwise purchase separately – any combination of lodging; sight-seeing, attractions, meals, entertainment, car rental, and transportation by air, motor coach, rail, or even private vehicle. A package tour may include more than one destination.

personal goodwill the value of profit generated over and above market expectations which would be extinguished upon sale of the specialized trading property, together with those financial factors related specifically to the current operator of the business, such as taxation, depreciation policy, borrowing costs and capital invested in the business.

pipeline study a survey of proposed hotel developments. The most famous is the W Hospitality Group Pipeline Study of new hotels being developed across Africa.

plate service a variation of table service: basic service style in which fully cooked menu items are individually portioned, plated (put on plates) in the kitchen, and carried to each guest directly.

platter service a table service style in which servers carry platters of fully cooked food to the dining room, present them to the guest for approval, and then serve the food.

PMG (price match guarantee) The promise that hotels or OTAs will offer the lowest rates or match the lowest rate available across any channel for the same product.

PMS (property management system) an application used by the hotel to control onsite property activities such as check in/out, folios, guest profiles, room status, requests, etc.

POM property operations and maintenance.

POMEC property operations, maintenance and energy costs.

portion cost the standard food cost for an item that is sold as a single menu selection. The portion cost indicates the cost incurred by preparing one portion of the menu item according to its standard recipe.

POS (point of sale system) a computerized system that retail outlets such as restaurants, gift shops, etc., enter orders and maintains various accounting information. The POS generally interfaces with the property management system (PMS).

PPPN per person, per night.

premium the price paid for a lease, in the open market, where one tenant assigns its interest to another, replacement tenant. *See also* reverse premium.

prime investment a property investment regarded as the best in its class and location, determined by occupier and investor sentiment. Typically, a modern or recently refurbished building, finished to a high specification, well-situated in a commercially strong location and let to a strong covenant.

private treaty the most common form of buying/selling a property, involving a binding private contract for sale between the parties. A sale of a property or investment opportunity by 'private treaty' allows the vendor more control over the sale process and any specific conditions that apply. However, the completion of the sale can take longer than other routes, such as auction.

PRPN per room, per night.

purchase costs with property acquisitions, a prospective purchaser will normally incur acquisition or purchase costs, which relate to legal and surveyor fees, VAT and stamp duty.

R&M repairs and maintenance.

rack rate the standard or default rate for a room, before any discounts (for example, advance purchase discounts) are applied.

rack rented an historic term, still in common use in rent review clauses of modern leases, to the effect that the rent is at a full open market level. A 'rack' rent is one which has been 'stretched' (derived from the medieval torture instrument) to the full extent which could reasonably expected on an open market letting.

rate parity the strategy that all distribution channels of a hotel should reflect the same rate for the same conditions for a particular room type. Rate parity is often used to gain customer loyalty and encourages guests to book directly with the hotel.

Red Book The RICS Valuation – Professional Standards. This is a mandatory tool that valuers must adhere to, first published in 1974, and regularly updated.

Red Book valuation The RICS 'Red Book' contains rules and practice statements for all chartered surveyors who undertake asset and other formal types of valuation.

refurbishment the process of restoring, renovating or modernizing a hotels rooms or public areas to bring them up to a certain standard.

rent the consideration paid by the tenant to the landlord for the ability to occupy premises under a lease.

rent review a periodic review (usually five-yearly) of rent during the term of a lease. The vast majority of rent review clauses require the assessment of the open market, or rack rental value, at the review date, in accordance with specified terms, but some are geared to other factors, such as the movement in the RPI.

REO (reasonably efficient operator) a market-based concept whereby the potential purchaser, and thus the valuer, estimates the maintainable level of trade and the future profitability that can be achieved by a competent operator of the business conducted on the premises, acting in an efficient manner. The concept involves the trading potential rather than the level of trade under the existing ownership so it excludes personal goodwill.

repair covenants repair covenants are the contractual obligations in a lease which identify the landlord and the tenant's liabilities to repair.

restrictive covenant a covenant that imposes a restriction on the use of land to enable the value and enjoyment of the land to be preserved.

reverse premium the opposite of a premium payment on assignment; instead of the outgoing tenant (assignor) receiving a lump sum for the lease, it pays the replacement tenant (assignee) to take the lease on.

reversionary yield a measure of investment analysis showing the relationship between the capital cost of acquisition and the estimated rental receivable at the next lease event.

RevPAR (revenue per available room) one of the standard benchmarking measures in the hotel industry. However, it is not straightforward, as the revenue is related to room-generated revenue rather than total revenue. It is calculated by taking all rooms' revenue and dividing by the number of rooms. It can also be calculated by multiplying the ADR (or ARR) by the occupancy rate.

RevPOR (revenue per occupied room) calculated by taking the total daily revenue (including ancillary revenues) and dividing it by the total number of occupied rooms at the hotel.

RGI revenue generator index.

RICS Royal Institution of Chartered Surveyors.

RMS (revenue management system) a software application hotels use to control the supply and price of their inventory in order to achieve maximum revenue or profit, by managing availability, room types, stay patterns (future and historical), etc.

ROI return on investment.

room inventory the volume of rooms available to be sold.

room service the department within a food and beverage division that is responsible for delivering food or beverages to guests in their guestrooms. May also be responsible for preparing the food and beverages.

room type a room type represents some form of categorization, set, or collection of rooms with some common element at the hotel that must be managed for marketing purposes within the hotel. For example, a room type might be a suite or a single room with a double bed, poolside or ocean-side. A room can belong to multiple room types.

rooms division the largest, and usually most profitable, division in a hotel. It typically consists of four departments: front office, reservations, housekeeping, and uniformed service.

rooms' yield average revenue of all rooms, divided by the number of rooms in a hotel, divided by 365 nights. Also known as RevPAR.

RPI (retail price index) a measure that examines the weighted average of prices of a basket of consumer goods and services. The RPI is used to track price changes associated with the cost of living. In the commercial property market leases are often granted with rent reviews occurring by reference to the RPI (or occasionally the consumer prices index) (normally on an upwards-only basis). Unlike the CPI Index, the RPI includes mortgage payments. The RPI was first introduced in 1947 and was for many years, the government's main measure of inflation. In 2003 the government announced that the UK inflation target would be based on the CPI, replacing the RPI for this purpose.

S&M sales and marketing.

security of tenure the protection (security) afforded to tenants of commercial premises in the UK by Part II of the Landlord & Tenant Act 1954 at the end of their lease.

service charge (1) a service charge is payable by a tenant for services provided by the landlord for the repair and maintenance of common parts, such as lifts, reception areas and the external structure of the building. The service charge usually includes managing agent's fees. It is normally collected quarterly in advance, at the same time as the rent. A service charge payment can be capped in order to limit a tenant's financial liability. The RICS has developed a service charge code of practice to improve standards and promote consistency, fairness, transparency and best practice in the management and administration of commercial property service charges. (2) A percentage of the bill (usually 10 per cent to 20 per cent) added to the guest charge for distribution to service employees in lieu of direct tipping.

shell and core a specification for a property being leased to tenant; shell and core means constructed but not fitted-out.

short lead bookings made at short notice (e.g. on the day of arrival or within a few days of arrival).

snag list a list of problems/issues that need addressed (usually as a result of a new hotel launch).

soft launch/opening partial launch of a hotel property, perhaps at a reduced service level, usually to test the service offering prior to launching in earnest.

SOPU single owner per unit.

sous chef second-in-command after the head chef.

split service a food service method in which servers deliver courses separately. Split service helps maintain food quality and safety because each course can be portioned and served when it is ready, eliminating short-term holding in the kitchen.

subject to contract a provisional agreement for contracts to be exchanged where either party can withdraw from the transaction.

subletting takes place when a tenant grants a new lease for their property, or part thereof, to an alternative occupier, for a period less than the residue of the tenant's lease. The period of the subletting must be at least one day less than the unexpired period of the superior lease. If a tenant attempts to sublet the property for a period equal to, or more than, the unexpired period of their own lease; this operates as an assignment of the term and not as a subletting.

sustainable tourism generally refers to environmentally conscious hoteliers/guests. They may request details of the hotel's carbon footprint etc. Some corporates may select a hotel based on its sustainable practice.

table d'hote a full-course meal with limited choice at a fixed price.

TCM traditional Chinese medicine.

third-party booking engine an internet site that provides a booking engine where a traveller can search a large number of lodging facilities for availability and reserve a room. The lodging facilities are not affiliated with the site and pays a fee for the business that the third party site generates. Examples of third party sites include: www.hotels.com, www.bookings.com, www.priceline.com.

TRevPAR (total revenue per available room) calculated by taking the total revenue of a property (not just room revenue) and dividing by the total number of rooms available.

triple net income the net income from a property investment after deducting ground rent, non-recoverable expenditure and void holding costs.

turnover rent a method of calculating all or part of the rent of commercial premises, by reference to the tenant's turnover. Exact terms vary between leases, but usually this is based on a percentage of gross receipts. The tenant will typically pay a base rent 'on account', either based on the previous year's rent or a percentage (e.g. 80 per cent) of open market rental value, with a balancing charge at year end. Most common in 'destination retail' locations such as regional shopping centres, and the licensed and leisure sectors.

unconstrained demand the true demand for a for a hotel regardless of any capacity limitations.

upgrade process by which a guest is offered a better room than he/she booked.

upsell process by which a guest is offered (at a cost) additional services or upgrades (often at the point of purchase or upon arrival to the hotel).

USALI (Uniform System of Accounts for the Lodging Industry) the industry-standard way of reporting trading data.

user clause this is a contractual provision within a lease, that specifies the use, or uses to which a property may be put and the uses which are prohibited. The formal classification of 'uses' are set out in the Town and Country Planning (Use-Classes) Order 1987 as amended, which is a statutory instrument defining various use-classes. The terminology of a user clause contained within a lease is critical in determining whether the use specified is restrictive or open. This aspect can be very important when considering the open market rental value of a property.

VP (vacant possession) in the hotel industry VP means unencumbered by a lease or management agreement.

walk-in a guest that hasn't pre-booked, but simply walks in and reserves a room. Often they'll pay a higher rate (even rack rate) accordingly.

WBE web booking engine.

yield a measure of the return on an investment. A yield is the reciprocal of the multiplier that converts an income stream into a capital value.

yield management the process of understanding, anticipating and reacting to consumer behaviour to maximize revenue. Yield management is also referred to as revenue management.

YOY year-on-year.

YP (years purchase) this is a number that stabilised profit is multiplied by to arrive at a gross value. It is calculated by working out how many years (ignoring finance costs and the net present value of income) how long it would take for the purchaser to have repaid the purchase cost. For example, if a property was generating a profit of $8.4 m, and had been purchased for $110 m, that would equate to a YP of 13.09. This in turn can be used to demonstrate the capitalisation rate by dividing by 100. So in this case the cap rate was 7.63 per cent.

YTD year to date.

Index

Note: Page references in *italic* indicate figures, tables and boxes.